Using Powder Metallurgy in Design

Based on papers presented at the one-day Powder Metallurgy in Design for Wear, Corrosion, and Fatigue Resistance, held at the IMechE Headquarters, London, UK, on 17 September 1999.

IMechE Seminar Publication

I MECH E

Using Powder Metallurgy in Design
Wear, Corrosion, and Fatigue Resistance

Organized by
The Tribology Group of
The Institution of Mechanical Engineers

Co-sponsored by
The European Powder Metallurgy Association
The Institute of Corrosion

IMechE Seminar Publication 2000–5

Professional Engineering Publishing

Published by Professional Engineering Publishing Limited for The Institution of Mechanical Engineers, Bury St Edmunds and London, UK.

First Published 2000

ISBN 978-1-86058-303-2

A CIP catalogue record for this book is available from the British Library.

Related Titles of Interest

Title	Editor/Author	ISBN
Tribology of Magnetic Storage Systems	B Bhushan	1 86058 174 9
IMechE Engineers' Data Book	C Matthews	1 86058 175 7
An Introductory Guide to Industrial Tribology	J D Summers-Smith	0 85298 896 6
A Tribology Casebook	J D Summers-Smith	1 86058 041 6
New Directions in Tribology	IMechE	1 86058 099 8
World Tribology Congress	IMechE	1 86058 109 9

Contents

S649/001/99

Powder metallurgy (PM) – the process and possibilities

D WHITTAKER
Consultant, Wolverhampton, UK

1. THE PROCESS

The basic process for the manufacture of PM products involves two stages (Figure 1):

(a) The **compaction** of blends of powders, with suitable additives (e.g. pressing lubricants), to form the required shape. Compaction is most commonly carried out in rigid tooling (die and punches), but in some PM sectors, isostatic pressing in flexible moulds is also a well-established practice.

(b) **Sintering** of the compact to generate strength, integrity and an appropriate microstructure. Sintering is a controlled atmosphere heat treatment process carried out at a temperature below the melting point of the major constituent.

For some applications, this is the entire process. For many others, a range of post-sintering "secondary" operations may also be applied (see Figure1).

2. WHY USE PM?

There are two broad reasons for choosing to use PM:

(a) The process route, starting from a powder feedstock, allows the derivation of a characteristic which would be impossible (or at least very difficult) by a more conventional process i.e. **Product Uniqueness** is offered.

(b) PM is in competition with other potential manufacturing routes and wins the competition because it offers cost savings in terms of the finished product i.e. **Product Cost Effectiveness** is the selling point.

Judged on the basis of tonnage or numbers of components produced, the latter reason is dominant. However, there are several important PM application sectors driven by the former advantage.

3. APPLICATIONS BASED ON PRODUCT UNIQUENESS

(a) The production of **intimate mixtures of metals and/or non-metals, usually** regarded **as immiscible in conventional processing**, e.g. sintered friction materials, hard metals, Valve Seat Insert and Valve Guide materials.

(b) The manufacture of products with controlled **levels of porosity** e.g. PM self-lubricating bearings, sintered metal filter elements.

(c) The processing of **high melting point materials** e.g. refractory metals (W, Mo, Ta), ceramics.

(d) In certain circumstances, processing by PM can impart **superior physical properties** e.g. PM high speed steels, high performance permanent magnets, metal matrix composites.

4. APPLICATIONS BASED ON COST EFFECTIVENESS

(a) What makes PM cost effective?

Compared with other process routes, involving significant finish machining, PM draws its cost effectiveness from its high material utilisation and relatively low energy consumption (Figure 2). The cost effectiveness of PM for a particular application is enhanced by:

– **Intricacy of shape** – the more intricate the shape, the higher the cost of generating it by machining, the more likely that the net-shape PM process will offer cost advantages.

– **Close dimensional tolerance requirements** – dimensional control after sintering is generally around +/- 0.1 % in the plane perpendicular to the pressing direction and around +/- 0.5% in the pressing direction. A sizing or calibration operation after sintering can improve these figures by perhaps a factor of two.

– **High Production Volumes**

(b) Why high production volumes?

In common with other net shape forming processes, PM utilises high capital cost equipment and forming tooling which is relatively expensive but capable of producing several hundred thousand pieces before replacement. Consumable tooling costs and capital equipment depreciation charges need to be amortised over large numbers of products.

Also, tool changeover and, particularly, tool setting times can be long for complex parts. So, high volume batch production is desirable to minimise press downtime.

For most applications, there is a "break-even" point in terms of production quantities, below which it will be more cost-effective to machine the shape from the solid. This break-even point will vary by application, but is likely to be of the order of several thousand to a few tens of thousands.

5. HISTORICAL GROWTH OF PM STRUCTURAL PART APPLICATIONS

Before World War II, PM structural part applications were limited to plain bearings and bushes.

During the war years, the need to conserve raw materials and energy gave an extra impetus to the development of PM and the first shaped parts appeared.

Since then, PM has extended its business by tackling components with increasing degree of complexity and size (Figure 3). The PM structural parts business has now grown to a worldwide tonnage level of around 650 000 tonnes (see Figure 4 for split by geographical region).

Although there are other important end-user sectors, the automotive industry has emerged as the dominant market place for PM structural parts. Worldwide, around 80% (by weight) of all PM parts go into automotive applications (Figures 5 and 6).

PM parts can be found throughout the vehicle, but engine and transmission components account for over 70% of all automotive usage (Figure 7).

6. FUTURE GROWTH PROSPECTS

Continued growth of PM structural part applications will depend on extending capability against the factors generally perceived as limitations of the technology, i.e.:

- Shape – particularly in the pressing direction
- Size
- Strength

A series of studies throughout the 1990s has indicated that there remains considerable potential growth in the automotive components market – this market could be around 3 times larger than its present size.

The key to capturing this increased business has been identified as the PM industry's ability to persuade design engineers of PM products' performance capabilities and fitness for purpose in more exacting operating conditions.

Many material and process developments have appeared in recent years aimed at improving properties and performance of PM parts. Several of these are now making an impact in the market place.

However, the application of improved design methodologies is also appearing as an increasingly important contributor to persuading end-users of PM's capabilities in higher performance parts.

7. EXAMPLES OF THE CONTRIBUTION OF DESIGN ANALYSIS

(a) The technical viability of single press/sinter connecting rods (as opposed to Powder Forged rods) has now been demonstrated through the application of Finite Element Analysis (FEA). There is significant promise that such rods may be introduced in the next generation of automotive engines.

(b) FEA has also been used to model the influence of component geometry variations on the distribution of operating stresses in PM engine main bearing caps, as an aid to defining acceptable dimensional tolerance variations.

(c) FEA has also contributed to demonstrating the viability of improving the tooth bend fatigue strength and surface contact fatigue strength (pitting wear resistance) of PM gears by local surface densification. This approach is now offering products for automotive manual transmission applications.

(b) An alternative approach to introducing PM gears into higher performance applications has been to change the design of tooth profiles in order to reduce operating stresses.

8. CONCLUSIONS

1. The basic Powder Metallurgy (PM) process involves two stages – Compaction to form the required shape, followed by Sintering to provide strength and integrity. In many cases, post-sintering "secondary operations" may also be involved.
2. There are two major reasons for choosing PM – Product Uniqueness or Product Cost Effectiveness.
3. PM's cost effectiveness comes from its high material utilisation and low energy consumption.
4. Cost effectiveness is enhanced for the production of intricate close tolerance parts in high production quantities.
5. Markets for PM structural parts have developed historically mainly based on increased intricacy of shape and increased component size/weight.
6. The automotive industry has proved to be the most important market, consuming around 80% of PM's worldwide production (of around 650,000 tonnes of components).
7. Although there are PM applications elsewhere in the vehicle, automotive applications are predominantly in the engine and transmission.
8. The key to future growth in business is seen as tackling higher performance applications.
9. The use of modern design methodologies is making an increasing contribution to furthering this aim.

10. The technical viability of Pressed/Sintered PM Connecting Rods has now been demonstrated.

11. FEA has been utilised to define the dimensional tolerance requirements of PM engine main bearing caps.

12. Surface densified PM gears have now been developed for automotive manual gearbox applications, with FEA having been used to demonstrate that the approach is technically viable.

13. Design modifications to gear tooth profile have been shown to make undensified PM materials viable for transmission gear applications.

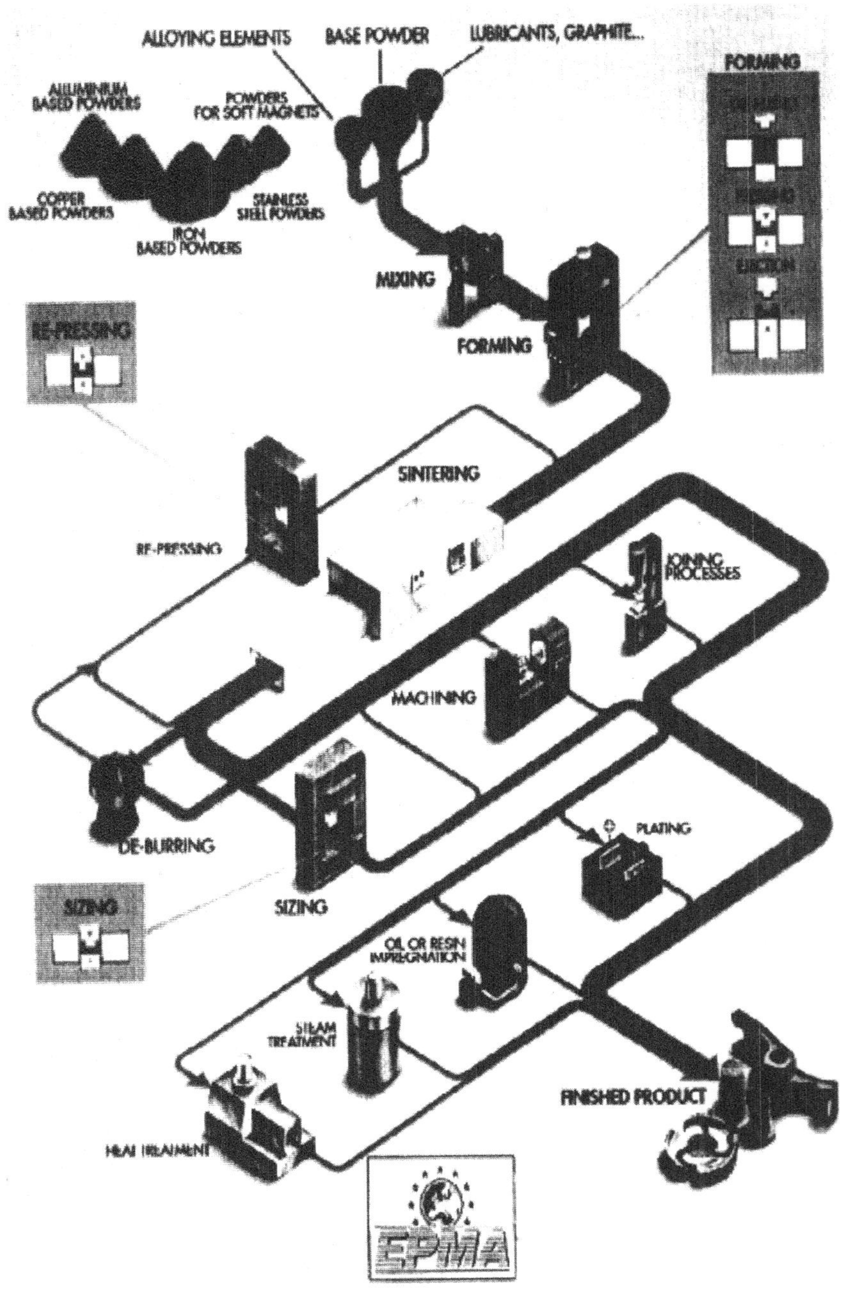

Figure 1. Powder Metallurgy – production cycle

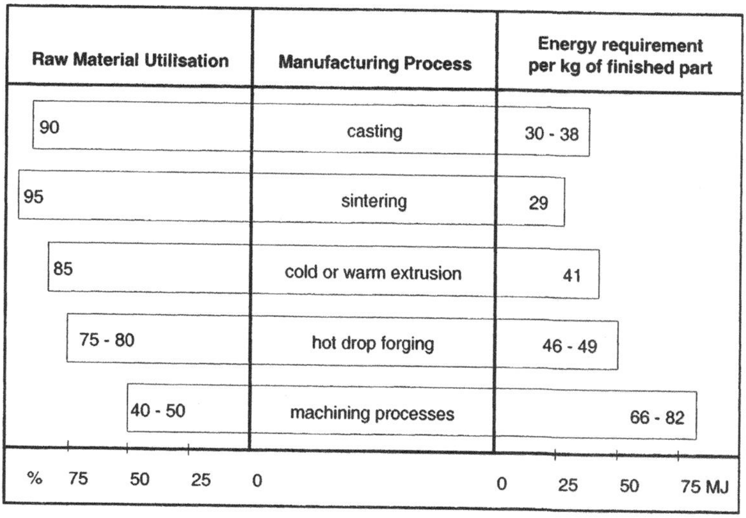

Figure 2. Why PM is cost effective – high material utilisation; low energy consumption

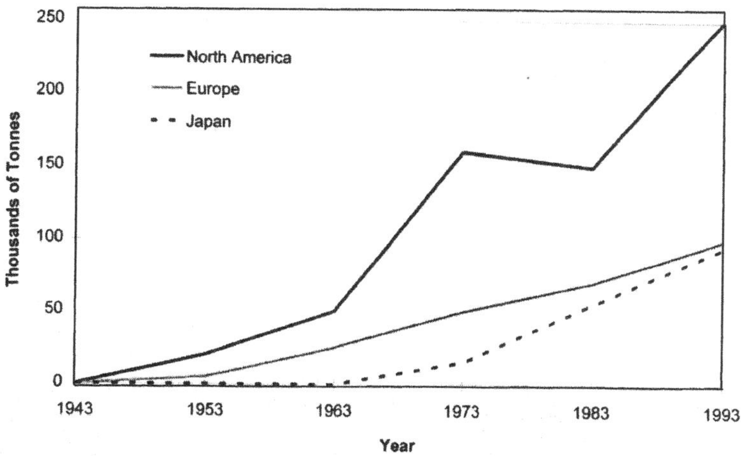

Figure 3. PM structural parts – global growth in sales

• USA	~ 335,000 Tonnes
• Europe (Western and Eastern)	~ 138,000 Tonnes
• Japan	~ 96,000 Tonnes
• Far East (excl. Japan)	~ 65,000 Tonnes
• South America	~ 13,000 Tonnes
WORLDWIDE TOTAL	**~ 650,000 Tonnes**

Figure 4. PM Structural Parts – Current Tonnage Levels

• Europe	~ 80%
• Japan	> 85%
• USA	~ 75%
WORLDWIDE	**~ 80%**

Figure 5. PM structural parts – automotive application focus

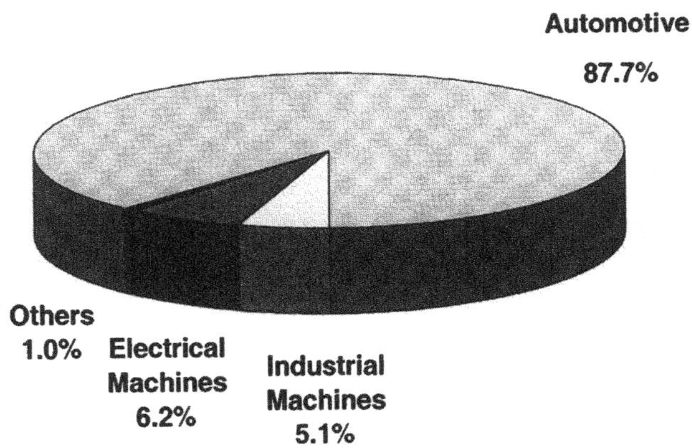

Figure 6. PM STRUCTURAL PART DEMAND JAPAN 1993

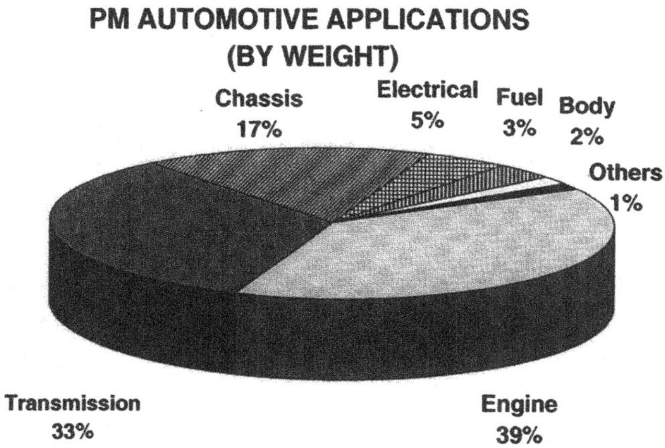

Figure 7. PM AUTOMOTIVE APPLICATIONS (BY WEIGHT)

Designing with powder metallurgy (PM) for adhesive wear resistance

I R WHITAKER
Federal-Mogul Sintered Products Limited, Coventry, UK

SYNOPSIS

Adhesive wear is said to occur when contacting surfaces are in relative motion resulting in transference of one material to another by solid-phase welding. Materials possessing self-lubricating properties are able to resist adhesive wear by the formation of thin surface films, which reduce intimate metal-to-metal contacts. Powder metal materials are produced with self-lubricating properties for operation in aggressive environments; this paper highlights a few applications where powder metallurgy, through design, is solving problems.

1. INTRODUCTION

This paper will attempt to illustrate how powder metal microstructures have evolved for wear applications, applications where adhesive wear can take place, rather than to present an analysis of adhesive wear resistance or the mechanism of wear. Adhesive wear can rarely be regarded in isolation from other wear mechanisms such as abrasive, erosion, fretting, rolling contact fatigue and chemical, except perhaps in the laboratory. Any analysis will tend to be specific to factors such as test apparatus, atmosphere/environment, counterface material and if lubricant is used, lubricant type, additives etc. The short literature review which follows is by no means exhaustive but introduces some of the work reported by other workers where the adhesive wear resistance of ferrous based powder metal materials has been evaluated.

Traditional metallurgy has provided the essential information on how to prevent adhesive wear; avoid sliding similar metals together and where possible lubricate the contact area[1]. Powder metallurgy can provide materials with composite microstructures where combinations of hard phase and soft phase reduce similar metal contact. Alternatively, liquid or solid lubricants can be incorporated into the matrix providing surface films which keep surfaces apart. However, during the material design process the adhesive wear properties obviously need to be considered along side other requirements placed on the component such as cost, ease of manufacture, machinability, corrosion or hot oxidation resistance, strength and toughness, making compromises inevitable.

There remains today a driving force to consider the powder metal route for many applications due to the economy of near net shape, and even net shape, manufacture. Property comparisons usually have to be undertaken between powder metal materials and traditional materials before designers will consider application testing. The literature contains many references to the wear properties of powder metal materials and comparisons are often made between powder metal materials and wrought/cast materials with the same composition. The data obtained from wear evaluation though is difficult to utilise due to the many application dependent variables, and designers faced with this scenario often resort to purpose built test apparatus specific to the application. Although the composition of a powder metal material can be matched to a wrought material, the majority of the powder metal materials manufactured contain inherent porosity which remains after sintering. Several authors have attempted to evaluate the role of the porosity on the wear characteristics of sintered steel[2,3] and it has been found that surface porosity can beneficially trap wear debris resulting in reduced wear rates.

The vast majority of wear mechanisms involve a counterface material which can contribute lesser or greater amounts to the wear process. The importance of the counterface material has been highlighted[4,5] and the transition to severe wear for sintered iron is moved to higher loads when the counterface material is changed from steel, to chromium, to grey cast iron. The graphite flakes in the grey cast iron tend to lubricate the contact interface. The transition is also shown to shift to higher loads by increasing the density of the sintered iron.

The nature of the powder metal process allows for the easy addition of other elements or compounds which may dissolve during sintering or remain at prior particle boundaries, generating microstructures quite different from those available traditionally. The lubricated adhesive wear resistance of common iron powder systems[6] has been investigated with respect to additions of carbon, phosphorous and copper. It was found that in general the adhesive wear resistance was improved with small additions of ferrophosphorous, graphite and copper, though graphite produced the most consistent results.

One unique feature of a powder metal component is the interconnected porosity which exists up to densities of approximately 7.2 Mg/m^3 in iron based components. Porous bronze bearings can provide self-lubricating properties when the pores are filled with oil[7]. The pin-on-disk adhesive wear characteristics[8] of sintered steel have been considered both with externally applied oil lubrication and with lubrication provided from within the pores following vacuum oil impregnation. No wear was observed with external lubrication and impregnated material gave comparable friction and wear behaviour. Impregnation with polymeric materials is also possible for specific applications[9]. The oil impregnation process is now routinely used during the manufacture of powder metal valve guides for automotive and heavy duty engines at Federal-Mogul Sintered Products Ltd[10] providing an interfacial oil film after periods of standing with the engine stopped.

Novel microstructures, different from those attainable by traditional routes, were developed for automotive exhaust valve seat inserts following the introduction of unleaded gasoline. It was observed during testing with unleaded fuel that exhaust valve seat inserts wore quite severely[11]. Further examination showed that the lead in the leaded fuel had provided a lead oxide barrier which prevented adhesive wear between the seating surfaces of the valve and valve seat insert[12]. Some manufacturers of powder metal valve seat inserts subsequently incorporated lead into the metal matrix to provide a surface film[13,14]; the lead was either admixed in the initial powder blend, or alternatively, infiltrated after sintering. Federal-Mogul Sintered Products Ltd. developed a range of materials based on high speed steel and a composite of high speed steel and iron[10,11]. These materials and their derivatives optionally contain molybdenum disulphide solid lubricant and are also infiltrated with copper alloy during sintering. A distribution of metallic sulphides is formed throughout the microstructure which provide surface films during interaction with the valve, thus reducing metal-to-metal contact. The adhesive wear resistance of composite microstructures has also been demonstrated in the laboratory[15,16,17] using hard phase additions to steel such as high speed steel and Triballoy. Attempts have been made to simulate valve seat insert wear in the laboratory using special purpose rigs[18,19,20], though the nature of the wear is complex involving adhesion, abrasion, corrosion, fretting and fatigue to greater or lesser extents in particular engines. In general, wear is found to increase with increasing time and load, but decrease with increasing temperature. Higher temperatures are thought to lead to the formation of an oxide film which prevents metal-to-metal contact and therefore reduces adhesive wear.

Novel powder metal microstructures have also been developed for other automotive applications which undergo adhesive wear. Cam follower wear[21] has been evaluated in terms of material properties and microstructure. Using a modified automotive cylinder head, wear was found to decrease with increasing macrohardness, bending strength, carbide hardness and the total fractional area of carbides. A distribution of fine globular carbides was preferred to a distribution of coarse angular carbides, a feature also observed in valve seat insert materials[22].

Research has been ongoing into the use of powder metal high speed steels for bearings in aeroengine gas turbines[23], components which have been traditionally made from forged M50 and T1 grades, with the possibility that the powder metal route might provide increased material utilisation and more uniform microstructure. Powder metal grades of both M50 and T1 were produced and compared to wrought material using a hot wear pin-on-disk apparatus. The sintered and wrought materials gave comparable wear performance. Above 400°C the wear rate of the molybdenum high speed steel decreased whereas that of the tungsten high speed steel remained constant due to a difference in the oxidation behaviour. Special powder metal grade steels incorporating alloy carbides have been developed for adhesive wear resistance[24] which incorporate 40 – 45 volume percent of TiC in a martensitic steel matrix. The TiC particles create an inert contact surface, when prepared correctly, thus eliminating metal-to-metal contact.

Valve seat inserts for heavy duty, turbocharged diesel and gas engines, particularly the inlet seat, can suffer from severe adhesive wear as there is no passage of fuel through the valve aperture during the inlet stroke to provide lubrication. A novel, full density M42 based high speed steel[25] was developed at Federal-Mogul Sintered Products Ltd. which has been evaluated in this application with considerable success. In addition, the incorporation of a solid lubricant through a reaction sintering route[26], originally developed for self-lubricating valve guides[27], has produced a material demonstrating remarkable wear resistance. The mechanism controlling the wear behaviour has not been analysed in detail, however other workers[28] investigating solid lubricant effects suggested that S catalysed the oxidation of iron in sintered Fe-Mo-S materials thus preventing metal-to-metal contact.

Solid lubricants have been investigated for wear resistant components in turbocharger applications such as wastegate bushes[29]. In the absence of surface films, elevated temperatures will promote metal-to-metal adhesion due to the increased mobility of the diffusing species. Diesel exhaust gas temperatures can typically reach 750 – 800°C and high alloy wrought stainless or cast steels containing 30 wt.% chromium have been traditionally used which possess the combined high temperature properties of stable oxide film formation and strength. However, as the benefits of turbocharging become more appealing to engine manufacturers, the complexity of turbocharger design increases. Powder metallurgy is proving to be very cost competitive in this arena through the successful application of microstructural design. By combining an austenitic stainless steel with a high alloy ferritic chromium steel and solid lubricant, a combination of property requirements have been satisfied, including high temperature adhesive wear resistance.

Adhesive wear properties have also been improved by the application of a steam treatment technique to generate an adherent layer of Fe_3O_4 on the surface of the powder metal part and within the interconnected porosity[30]. Sealing rings such as those used in shock absorbers are often steam treated prior to oil impregnation at Federal-Mogul Sintered Products Ltd. where enhanced wear resistance is demanded.

2. APPLICATIONS

2.1 Valve Guide

Materials for valve guide applications are designed to resist adhesive wear through the control of carbide distribution and morphology, oil impregnation and the incorporation of solid lubricant. Machinability is also a very important design criterion and the incorporation of carbides in to the microstructure is not desirable from this point of view. Therefore, materials have been developed according to the stress level and counterface material. Medium to high stress applications use a base iron powder containing phosphorous which is blended with graphite to produce carbides in a pearlite matrix, figure 1. A range of carbide morphologies can be produced depending on the sintering temperature, thus where high loads are experienced a higher proportion of surface carbides can be easily achieved for increased wear resistance. Copper is added for additional matrix strength and size control. Premium products are fully infiltrated with phosphor bronze for operation at high temperatures, the copper alloy providing high thermal conductivity. Low to medium stress applications use a base iron powder blended with graphite, molybdenum disulphide and calcium carbonate which produces a pearlite matrix with a fine dispersion of metal sulphides, figure 2. Grain boundary carbides are not formed.

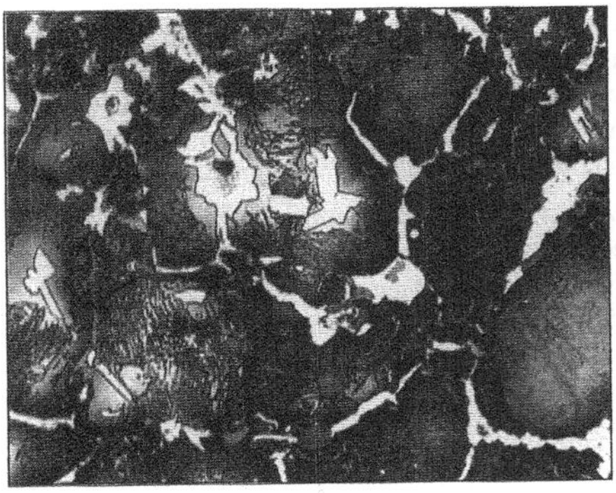

Figure 1. Carbides in a pearlite matrix x400

Figure. 2. Pearlite with metal sulphides x400

Materials are tested for wear resistance against a valve stem counterface using a custom built sliding wear apparatus. Usually the valve guide material is oil impregnated with high pressure turbine oil and the valve stem is coated with a thin film prior to testing. The friction force is continually measured throughout the 24 hour test. Figure 3 below shows that the surface film provided by the stem eventually breaks down resulting in metal-to-metal contact, whereas the oil impregnated valve guide continues to lubricate the interface. This test attempts to simulate cold start and marginally lubricated conditions in the engine and different valve guide materials and valve stem surfaces have been evaluated.

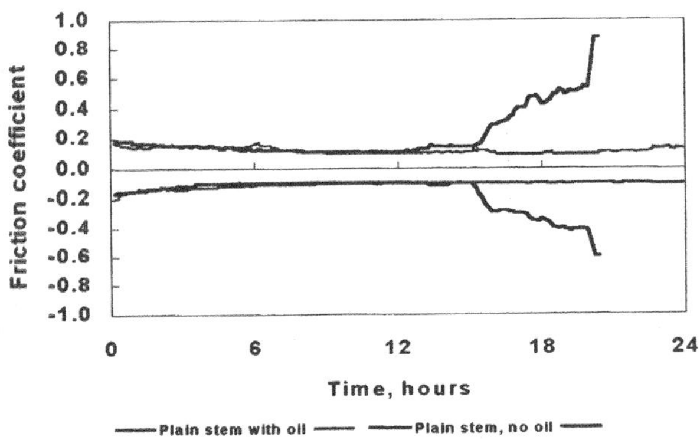

Figure 3. Effect of oil impregnation on friction coefficient.

2.2 Valve Seat Insert

Valve seat insert materials are designed to withstand high impact loads at high temperatures in aggressive environments, as well as adhesive and abrasive wear resistance. In the laboratory, materials can be ranked using the standard ASTM G77-93 block-on-ring test method prior to dynamometer engine testing. Wear resistance is again very dependent on the microstructure and alloy carbides provide high temperature wear resistance in a high strength tempered martensitic matrix. Exhaust valve seat inserts are therefore based on a high speed steel matrix for maximum wear resistance, figure 4, though hot work tool steel is used in less aggressive applications, figure 5.

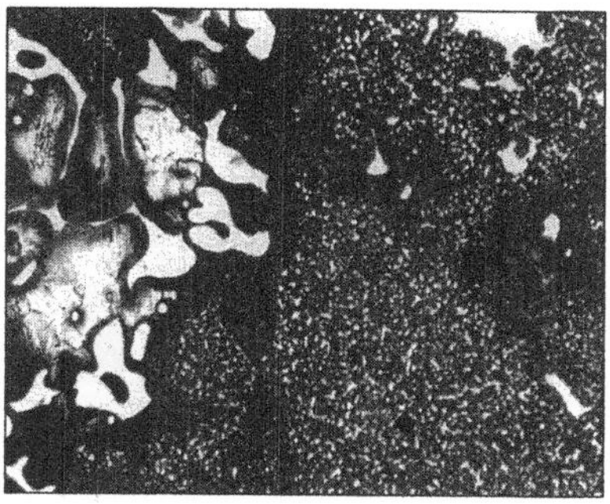

Figure 4. High speed steel base x400

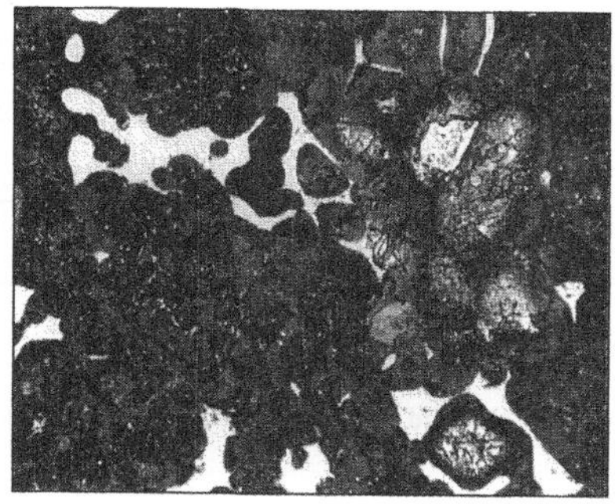

Figure 5. Hot work tool steel base x400

Often, exhaust valve seat inserts are infiltrated with copper alloy in order to increase the thermal conductivity, machinability and wear resistance, and the steel matrix can be diluted with up to 50% iron for lower duty applications. The copper alloy can contribute to the overall wear resistance through the formation of a copper oxide surface film at high operating temperatures. The comparative wear resistance of high speed steel and hot work tool steel materials is illustrated in figure 6. It is likely that the alloy carbides in the high speed steel, which are larger than those in the hot work tool steel, control the wear properties.

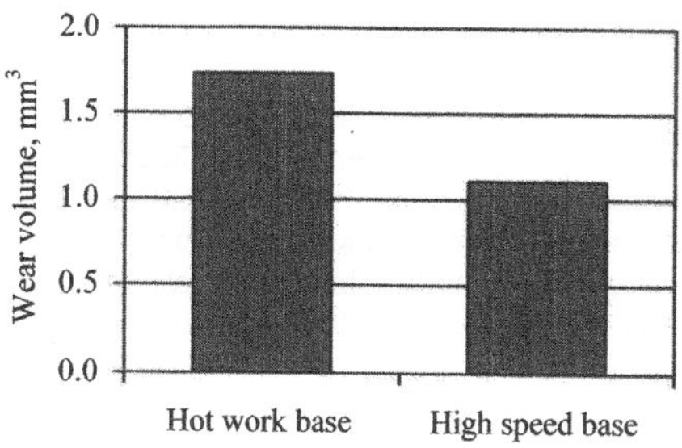

Figure 6. Block-on-ring wear resistance.

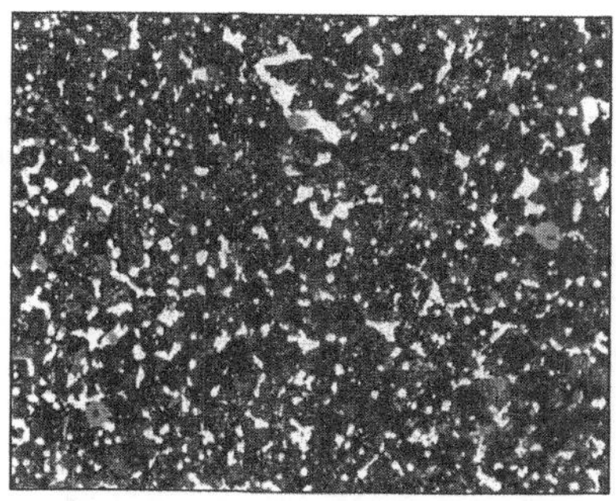

Figure 7. Full density high speed steel x400

Inlet valve seat insert materials for port injection gasoline engines are usually made from low duty pearlitic materials as adhesive wear is prevented to a large extent by the flow of the fuel mixture which lubricates the contact face. However, direct injection diesel engines require a higher degree of inlet valve seat insert wear resistance due to the dry nature of the inlet gas. Full density high speed steels have been successfully employed in heavy duty turbocharged diesel engines where alloy carbides resist wear, illustrated in the microstructure in figure 7. In addition, the incorporation of a solid lubricant has proved to be very beneficial in some applications, particularly where the valve material compatibility has not been optimised. Figure 8 shows the average combined valve and valve seat insert wear in a 6 litre, 6 cylinder, turbocharged diesel engine after 500 hours endurance testing. The solid lubricant has reduced the wear significantly.

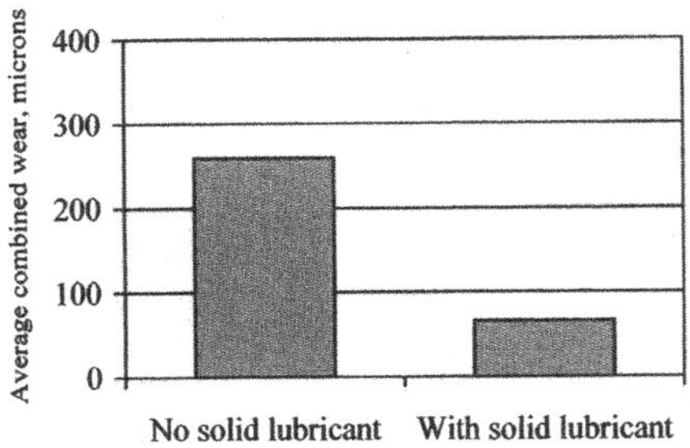

Figure 8. Combined vsi and valve wear.

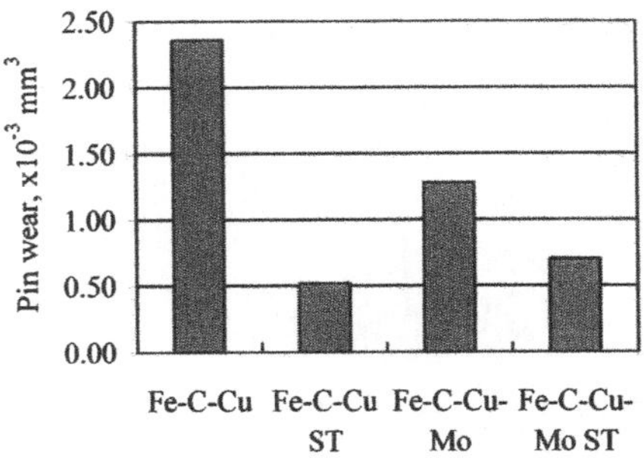

Figure 9. Pin on disk wear.

2.3 Sealing Rings

Shock absorbers for car suspension systems often contain sealing rings which control the flow of oil around the piston and can therefore greatly affect the damping efficiency of the shock absorber unit. Wear resistant materials are required which are compatible with the shock absorber tube. Powder metallurgy rings are made with a pearlitic matrix which is optionally steam treated prior to oil impregnation during the final stages of manufacture. Lubricated, reciprocating pin-on-disk tests have been carried out which demonstrate the improvement in wear following steam treatment. The results are shown in figure 9 for materials based on Fe-C-Cu and Fe-C-Cu-Mo where ST denotes steam treatment.

2.4 Turbocharger Bush

Bushing materials have been developed with 12 and 20 wt.% chromium in a wear resistant steel matrix. These materials have been used for high temperature reciprocating wear applications eg. turbocharger wastegate bushes, where temperatures can reach up to 700°C in exhaust gas atmospheres, wear resistance being provided by a combination of alloy carbides, solid lubricant and surface oxide. A custom built reciprocating wear rig was designed and built to evaluate material performance in diesel exhaust gas at 600°C[29]. The 20 wt.% chromium material was tested with a bearing pressure of 0.1 MPa, covering $132x10^3$ cycles in 110 hours and no wear was observed on either the bush or the shaft. Subsequently, the material has been adopted commercially under similar operating conditions. The microstructure illustrated in figure 10 shows a distribution of alloy carbides in a ferrite phase together with austenite prior particles. Small grey sulphide particles are distributed throughout.

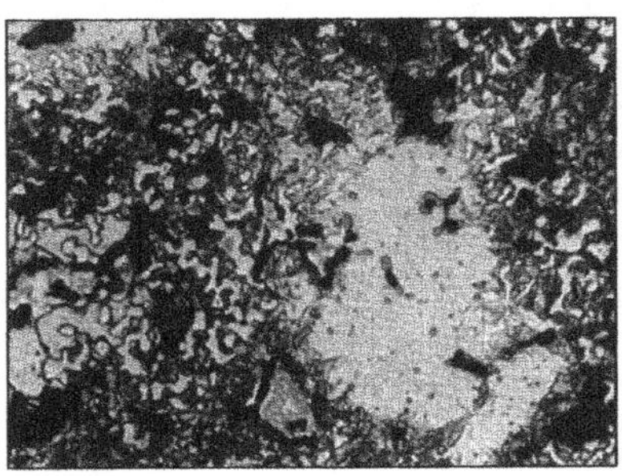

Figure 10. High temperature bush material x400

3. SUMMARY

The powder metallurgy process provides a technique for creating novel, wear resistant microstructures not attainable by any other technique. Adhesive wear can occur in many applications and several high volume niche products have been described. Although adhesive wear resistance is a design criterion, it is often not the only one and many other factors influence the final product.

4. REFERENCES

[1] K.C. Ludema, *ASM Handbook*, volume 18, p241
[2] B. Dubrujeaud, M. Vardavoulias and M. Jeandin, *Wear*, 174 (1994) 155 - 161
[3] E.R. Leheup, D. Zhang and J.R. Moon, *Wear*, 176 (1994) 121 - 130
[4] T.S. Eyre, *Powder Metallurgy*, 24 (1981) 57 - 63
[5] P. Kjeldsteen, *Modern Developments in Powder Metallurgy*, Proceedings of the 1984 International Powder Metallurgy Conference, Canada, pp467 - 482
[6] A.E. Miller and F.J. Perricelli, *Progress in Powder Metallurgy*, 33 (1977) 147 - 156
[7] V.T. Morgan, Perspectives in Powder Metallurgy: 4, Chapter 15, Plenum Press 1970.
[8] A.I. Taskinen and M.H. Tikkanen, *Modern Developments in Powder Metallurgy*, Proceedings of the 1984 International Powder Metallurgy Conference, Canada, pp545 - 567
[9] V. D'Agostino, V. Niola and G. Caporiccio, *Tribology International*, 21 (1988) 105 - 108
[10] C.G. Purnell, H.A. Brownlie and P. Maulik, Proceedings of the 1993 Powder Metallurgy World Congress, Kyoto, Japan, pp525 - 528
[11] M.S. Lane and P. Smith, SAE 820233
[12] C.S. Weaver, SAE 860090.
[13] A. Fujiki and M Kano, SAE 920570
[14] H. Danninger,E. Wolfsgruber and K. Dalal, PM94 Powder Metallurgy World Congress, Paris, France 1994, volume 2 pp875 - 878
[15] C.B. Cameron, R.A. Hoffman and R.W. Poskitt, Progress in Powder Metallurgy 1975, volume 31, pp41 - 51
[16] V. Manolache, D. Duda and W.J. Huppmann, *Powder Metallurgy*, 1981, no. 4 pp185 - 190
[17] P.D. Nurthen et al, Advances in Powder Metallurgy 1991, volume 5, pp135 - 149
[18] C.M. Hofmann, D.R. Jones and W. Neumann, SAE 860150
[19] Y.S. Wang, S. Narasimhan, J.M. Larson, J.E. Larson and G.C. Barber, *Wear*, 201 (1996) 15 - 25
[20] X. Liang, G. Strong, D. Eickmeyer and K. Myers, SAE 1999-01-1319
[21] M. Kato and Y. Kimura, *Wear*, 162 – 164 (1993) 897 - 905
[22] M. Onoda, N. Kuroishi and N. Motooka, SAE 880668
[23] T.H.C. Childs and A. Mimaroglu, *Wear*, 162 – 164 (1993) 890 - 896
[24] J.L. Ellis, *Powder Metallurgy International*, 16 (1984) 53 - 55
[25] I.R. Whitaker, R. Wyndow and C.G. Purnell, PM98 World Congress, Granada, Spain, 1998, volume 2 pp302 -307
[26] H.A. Brownlie and C.G. Purnell, PM94 World Congress, Paris, France, 1994, pp891 - 894
[27] P. Maulik and I.R. Whitaker, Powder Metallurgy in Automotive Applications, Proceedings of the International Conference on Powder Metallurgy for Automotive Components PM-97, February 1997, New Delhi, India, pp283 - 293
[28] Z. Guowei, O. Jinlin, M. Xiukun, M. Li, Q. Shangkui and Z. Fu, *Wear*, 162 – 164 (1993) 450 - 457
[29] I.R. Whitaker, P. Maulik and C.G. Purnell, PM98 World Congress, Granada, Spain, 1998, volume 4 pp259 - 264
[30] A. Molinari and G. Straffelini, *Wear*, 181 – 183 (1995) 334 - 341

S649/003/99

Designing with tungsten carbide for erosive/corrosive applications

R COOPER
Boart GmbH, Burghaun, Germany (formerly of Boart Longyear, Springs, South Africa)
D McHATTIE
Boart Longyear, Alcester, UK

SYNOPSIS

Hardmetal components have been used successfully for many years in erosive, and more recently in corrosive applications. Hardmetals can be tailored to give a wide range of toughness and wear resistance by varying the cobalt binder content and tungsten carbide grain size. However, to take full advantage of its unique properties, components must be designed for the material. It is not sufficient, as has happened many times in the past, to substitute hardmetal for a component which has initially been designed in more conventional engineering materials, such as hardened steel. As hardmetals have a relatively low fracture toughness and tensile strength, design has to take these factors into consideration. When successfully applied, hardmetal will provide dramatic improvements in erosive and corrosive wear resistance compared with most materials.

INTRODUCTION

Erosion has been defined as "the progressive loss of original material from a solid surface due to the mechanical interactions between the surface and a fluid or impinging liquid or solid particles". The degradation of material due to solid particle erosion poses a severe financial problem to industry, and the occurrence of erosion has been reported in many diverse fields, such as coal gasification plants, pipelines conveying solid particles and oilfield pumps and pipelines. In some cases, the effects of erosion are magnified by the corrosive nature of the fluid transporting the erodant particles.

The success of hardmetal in combating erosion and corrosion is due to the fact that it is a composite material, consisting of a hard, wear resisting phase, typically tungsten carbide (WC) particles, bonded together with a ductile binder, typically cobalt or a combination of cobalt, nickel and chromium. By altering the WC content and grain size, a wide range of properties can be obtained.

PROPERTIES OF HARDMETALS

Due to the refractory nature of WC, hardmetal must be produced by the powder metallurgy route, where the WC is milled with the binder, pressing or fabricating the component, followed by sintering, which melts the binder and subsequently bonds the WC particles during cooling down from the liquid phase. Cobalt, or mixtures of cobalt and other metals, are ideal as binders as they have excellent wetting and adhesion, together with good mechanical properties.

The properties of hardmetal are strongly dependent on microstructure, which in turn is determined by binder content and the WC grain size. This is shown in **Figure 1**, where the hardness decreases with increasing cobalt content and increasing WC grain size. On this figure, a 1 micron grain size WC bonded with 6% cobalt gives a hardness of 92 HRa, compared with the hardness of 82,5 HRa with a 5 micron WC grain size and 25% binder. It can be seen, therefore, that composition can easily be tailored to give the required hardness. Changes in WC grain size and binder content can be related by a quantity called binder mean free path, which is a measure of the volume percent of binder, mean WC grain size and the number of contiguous WC/WC and non-contiguous WC/Co boundaries. The relationship between transverse rupture strength (TRS), which is basically the tensile strength of the material in bending, and mean free path is shown in **Figure 2**. The 6% alloy shows an increase in TRS with an increase in mean free path. At the lower values of mean free path there are more WC/WC boundaries, which are weaker than the WC/Co boundaries, so strength will increase with increasing mean free path. For the higher binder contents, the strength increases with increasing mean free path, for the same reasons as with the 6% binder hardmetal, until a maximum value of TRS is reached, where there are an optimum number of WC/WC and WC/Co interfaces. Although binder content and mean free path are closely related, the maximum TRS increases primarily with an increase in binder content, which is considered to be due to the plasticity of the binder acting as a crack arrestor when fracture commences. At higher values of mean free path, strength decreases as the binder interlayers become thicker and strength is then more determined by the volume fraction of the binder phase.

As with all engineering materials there is an inverse relationship between toughness and strength. For many years the TRS value has been used as an indicator of toughness but equipment is now available for measuring the true fracture toughness of hardmetal. Whereas there is an increase in TRS with increasing binder content up to a maximum, and then a decrease, K1c increases linearly with binder content up to about 20%, after which the K1c values plateau. **Figure 3** shows the schematic relationships among compressive strength, TRS and hardness, with increasing binder content. It is apparent that hardmetal has a very high compressive strength and hardness, especially at lower binder contents, but the TRS is considerably lower. As binder content increases, hardness and strength decrease but TRS increases. The K1c values would be similar to the TRS curve, following the dotted line at higher binder contents. In attempting to capitalise on the high hardness and compressive strength of hardmetal, the limiting factor will be TRS and toughness. Component design must, therefore, minimise tensile stresses.

EROSION OF HARDMETAL

The useful life of a wear resistant part, irrespective of material, will depend not only on the material but on the environment in which it operates, that is, on the erodant slurry. Most research work has been carried out with silicon carbide grit or silica sand entrained in an air or liquid slurry stream under standard conditions, usually far more severe than in actual service as research results need to be obtained quickly. The designer has to consider material selection and design in terms of the erosive environment, hence the importance of its characteristics on erosion rate. Some empirical information for the erosion of engineeringly brittle materials is as follows:

- The degree of erosion increases with erodant particle size, up to a critical value which is dependent on the velocity of the slurry stream. As the velocity increases, so the critical size of the eroding particle increases, giving a synergistic relationship between particle size and velocity.

- Erosion rate increases with particle angularity, which is generally directly related to its hardness.

- Erosion rate also increases with increasing concentration of particles in the slurry, up to a certain limit, after which the erosion rate decreases. At high slurry concentrations, there is increasing interference between impinging and reflected particles near the surface of the body. The interference results in fewer strikes per unit time and decreased strike energy.

- Erosion rate increases as the incident angle of impingement of the slurry stream to the component increases, being a minimum in laminar flow conditions, i.e. 0 deg., and a maximum at an impingement angle of 90 deg.

Material loss by erosion has been studied extensively and some of the models are presented in **Figure 4** *Refs. 1-6.* The researchers examined the influence of the erodent on the Cobalt matrix and WC grains, each column shows their principal findings. Ball and Paterson's model is shown schematically in **Figure 5**, where a maximum erosive loss occurs at approx. 10% cobalt, followed by a drop to a minimum at approx. 20% cobalt, then followed by a gradual increase.

At lower binder contents, erosive loss is controlled by the contiguity of the carbide skeleton. At high contiguities, the load of the impacting particle is spread across the rigid skeleton, resulting in little grain fracture.

The wear rate is controlled by the extrusion of binder from between the carbide skeleton, leading eventually to lack of support of grains and their removal from the surface. This is shown in **Figure 6,** from *Ref. 8.* As contiguity decreases, that is, when binder content and/or carbide grain size increases, a weaker structure is formed due to fewer overlapping grains and more grains just touching. When a particle impacts an area where grains are just touching, a stress concentration is set up resulting in carbide grain fracture, as shown in **Figure 7.** An alloy containing approx. 10% cobalt appears to have the largest proportion of contacting WC grains, hence the highest wear rate.

Between 10 and 20% cobalt, the contiguity decreases further and there are fewer overlapping and contacting grains, so the likelihood of erosive particles impacting these boundaries reduces, hence the reduction in material loss. The wear rate reduces until an optimum of hardness, determined by the relative number of WC/WC and WC/WC-Co interfaces, and toughness, determined by the binder content, is reached. This results in a minimum of erosive loss.

At higher binder contents, an "embedded" structure of the WC grains in the cobalt predominates, so the wear mechanism changes to one dominated by the ductility of the binder. This is shown in **Figure 8**, where the impacting particle actually cuts through the binder, causing the debris to pile up. This debris would easily be removed on subsequent impacts. The wear rate will, therefore, increase with cobalt content.

An important factor governing wear rate is the angle at which the erosive particle stream impacts the component surface. **Figure 9**, from *Ref. 8* shows erosion rate as a function of cobalt content for various impacting angles. At low impingement angles, where the flow of erodant is approximately parallel to the component surface, as would occur in sand blast nozzles or pipes under laminar flow conditions, cobalt content has little effect on erosion rate. As the angle of incidence increases to 30 deg. the erosion rate increases with increasing cobalt, but the plots still remain relatively flat. The wear mechanisms already described will occur and their severity increasing with increasing impact angle. At incidence angles between 45 and 90 deg. a dramatic change occurs as there is now a peak in erosion rate at 10–11% cobalt, irrespective of incidence angle, the absolute erosion rate increasing with impact angle. This ties in with Ball and Paterson's model as in alloys with less than 10–11% cobalt a rigid carbide skeleton is present and the load due to an impacting particle is transferred across the rigid skeleton. Grain fracture does not occur but extrusion of the cobalt and its subsequent wear will lead to WC grain loss from the surface. With increasing cobalt, the skeleton becomes less rigid, as described, until maximum wear occurs at 10–11% cobalt. Thereafter, erosive wear decreases, but in this case there is no minimum at approx. 20% cobalt as shown by Ball and Paterson.

Ball and Paterson's model, expanded by Pennefather shows that at cobalt contents less than 10%, erosion rate is controlled by the "brittle" WC skeleton, characterised by the peak in erosion rate, but above this, erosion rate is controlled by the "ductile" binder phase, characterised by a more linear increase in erosion rate with cobalt content. From this it can be implied that at low impact angles the "ductile" mode predominates, irrespective of cobalt content, as there is a linear relationship between erosion rate and impact angle, irrespective of cobalt content. At higher impact angles, the "brittle" mode is predominant, leading to the peak in erosion rate.

CORROSION–EROSION OF HARDMETALS

Although the WC skeleton in hardmetals is inert to most corrosive media, the cobalt binder is not. For example, in a typical industrial grade containing 10% cobalt with a mean WC grain size of 2 microns, corrosion rate in 0,5N NaCI is 0.07 mg/cm2/day at ambient temperature and 0.12 in a boiling solution. In 6N HCl, the ambient corrosion rate is 1,0 mg/cm2/day compared with 15 in a boiling solution. These corrosion rates were obtained by leaving weighed specimens in the corrosive fluid for 28 days and then calculating the weight loss and corrosion rate. Anodic polarisation curves from *Ref. 9* showed an increase in passivation behaviour, decrease in corrosion potential and drop in current density with an increase in nickel content. The passive film, and its resistance to the impact of the eroding particles will be of importance. In corrosive media, the volume loss due to corrosion will be additive to that of erosion, as shown in **Figure 10,** from *Ref. 9*. The effect of a sea water slurry stream compared with tap water is evident. In an attempt to reduce the influence of corrosion on the overall wear rate, alternative binder systems have been produced. Typical results are shown in **Figure 11**. The compositions of the grades were:

Boart Longyear C6 - 94,0% of, 6% Co
Boart Longyear P6 - 94,0% WC , 3,0% Ni , 2,4% Co , 0,6% Cr
Boart Longyear V7 - 94,0% WC , 5,67% Ni , 0,33% Cr
Boart Longyear V6 - 94,0% WC , 6,0% Ni

It can be seen from **Figure 11** that the P6 grade shows the lowest erosion rate in either salt water or tap water and that the 6,0% nickel binder V6 is far worse. Although nickel has a greater corrosion resistance than cobalt in most media, its mechanical properties as a binder are worse due to its lower hardness, hence poorer wear resistance. Its passive film is readily eroded by a slurry stream. A nickel binder can be dramatically improved by chromium additions, which form a solid solution with the nickel and causes some WC grain refinement, increasing its hardness from 1448 VPN for V6 grade to 1674 VPN for V7. This hardness increase, rather than the presence of a passive film will, most likely, be predominant in governing erosive–corrosive wear rate. The P6 grade was developed, not only to improve corrosion resistance by adding nickel, but also to develop a work hardening binder. Brabyn *et al. (Ref. 10)* shows that the work hardening of the binder increases to a maximum at a composition of 70% cobalt and 30% nickel. The work hardening during local plastic deformation of the binder will, therefore, make it more difficult to extrude binder out of the WC skeleton in lower binder content alloys, and more difficult to gouge out material from a components surface in higher binder content alloys. The Boart Longyear grade P6, with its Ni/Co/Cr work hardening binder, and WC grain refinement through the Cr additions has therefore, the lowest erosion rate, in both tap and sea water, of all the grades tested by Wentzel *et al.*, reported in *Ref. 9.*

DESIGNING WITH HARDMETAL IN EROSIVE–CORROSIVE ENVIRONMENTS

In cases where a customer is dissatisfied with the performance of a conventional material, such as hardened steel in a wear resisting application, it is natural for him to think of alternative materials, such as hardmetal. In many cases, a direct design substitution is made, the customer possibly not being aware of the mechanical properties of hardmetal which can preclude some design features. Some design factors, in terms of acceptable and unacceptable features, are shown in **Figures 12** and **13**. Often, the fact that components have not been designed for hardmetal has resulted in catastrophic failure. For example, the choice of a hard, erosion resistant hardmetal grade for highly stressed components containing stress concentrations, such as holes, sharp corners, or sharp changes of section, which may not be problematical in a ductile material, are likely to be disastrous.

The component manufacturer should work closely with the hardmetal supplier to optimise both design for hardmetal and grade selection for the given application. Conventional stress analysis is used to determine the principal tensile stresses, which will govern failure. A useful rule of thumb criterion is that the tensile principal stress should not exceed 20 to 25% of the hardmetal TRS. As hardmetal has negligible plasticity in the engineering sense, stress overloads are difficult to accommodate and can cause rapid fracture, hence the choice of a relatively low allowable stress. If the design is acceptable, the environment has to be considered. Corrosive applications will ideally require a nickel or cobalt/nickel grade, although in the past success has been obtained with cobalt bonded hardmetal. A highly corrosive-erosive environment will necessitate a cobalt/nickel/chromium grade.

A word of caution is necessary, as for equal binder contents, these grades have a lower fracture toughness than cobalt bonded grades. K1c can be up to 2 Mpa.m1/2 lower. Wherever possible, grades with a higher, rather than a lower binder content should be selected, especially in regimes where the angle of incidence of the erodant stream approaches 90 degrees.

Examination of eroded surfaces on conventional components is valuable in determining the type of wear occurring. Smooth flow lines demarcating the flow of fluid are typical of "ductile" type wear and can allow the use of hard, wear resisting grades of hardmetal, typically 6–10% cobalt. Surfaces which are characterised more by overall roughness, such as seen is heavy shotblasting are typically caused by "impact' type wear. In these cases higher cobalt contents, usually 12–20% are necessary to move away from the peak in wear occurring at high impact angles. This assumes that whichever grade is selected will also withstand normal operating stresses.

Some examples of where hardmetal has substituted for conventional materials to give dramatic improvements in life are shown:

OILFIELD PUMP DISCHARGE COVER

A similar application is the pump discharge cover, shown in **Figure 14** where the sand entrained in the fluid stream has caused severe "ductile" wear of the wall. The typical life of a steel pump cover was 1500 hours, but with a hardmetal insert, after 9,000 hours running there has been virtually no wear on the cover and the expected life with the Hardmetal insert fitted is 30,000 hours.

The pump body contains 3 holes and these had to be reproduced in the hardmetal insert. In addition, these holes intersected the 90 degree angle inside the insert. No design change could be made to eliminate these undesirable features, which precluded the use of a grade containing less than 10% cobalt. In addition, the precision holes in the hardmetal had to be produced by spark erosion which leaves a microcracked surface. To remove this, the holes were honed and the edges of the holes were radiussed to remove sharp edges, which could act as stress concentrations.

COAL GASIFICATION PUMP

In this application, the pump operates at 28 bar pressure delivering a gas liquor containing ash, coal dust and tar. A steel bodied pump, shown in **Figure 15** typically lasted 3 months, but the use of a hardmetal sleeve showed negligible wear after 22 months in service. Again, in this case, the wear was "ductile", allowing the use of hardmetal with a low binder content of 6%. Shrink fitting of the insert was also advantageous.

WEAR PLATES

These are commonly used to protect less wear resistant materials, **Figure 16**, typical applications being mill liners, brick moulds, and sharp pipe bends. The plates can be subjected to very high impact loads, often at impingement angles approaching 90 degrees, hence resulting in "brittle" type wear. Grade selection must consider the high impact forces and the high impact angle of the slurry. For these reasons grades containing up to 20% binder are common. At 20% binder, the toughness will be high, hence to some extent mitigating stress overloads, and also the higher binder content is well away from the peak erosion region, occurring at 10–11% binder, at impingement angles approaching 90 degrees.

CONCLUSIONS

This paper has attempted to explain the erosive–corrosive wear mechanisms and their relationship to material properties, hence allowing the mitigation of wear and minimising the probability of catastrophic failure in service by fracture. Of importance is the holistic approach to the use of hardmetal in terms of design and grade selection.

Manufacturers of components which have to resist erosive–corrosive environments should never merely substitute hardmetal for a component designed for conventional, ductile engineering materials. The Hardmetal manufacturer must be involved in all stages of design for a given application to ensure optimum life of the component.

REFERENCES

1. Blomberry R.I. *et al.* – "Similarities in the Mechanism of Wear of WC-Co Tools in Rock and Metal Cutting". Wear, 1974, Vol. 27, pages 383–390.

2. Hansen J.S. *et al.* – "Erosion Testing of Potential Valve Materials for Coal Gasification Systems". Bureau of Mines Report 1979, R.I. 8335.

3. Conrad H. *et al.* – "Erosion of Sintered WC-Co Alloys". Proceedings of the International Conference on Recent Developments in Special Steels and Hard Materials". Pretoria, South Africa, 1982, pages 423–429.

4. Wright I.G. *et al.* – "Slurry Erosion of WC-Co Cermets and its Relationship to Materials' Properties". Proceedings of the 6th. International Conference on Erosion by Liquid and Solid Impacts". 1983, pages 63.1–63.8.

5. Shetty, D.K. *et al.* – "Slurry Erosion of WC-Co Cermets and Ceramics". ASLE Transactions, 1984, Vol. 28, pages 123–133.

6. Ball A., Paterson A.W., – "Microstructural Design of Erosion Resistant Hardmetals". Proceedings of the International Conference on Recent Developments in Special Steels and Hard Materials, 1985, Rhodes, pages 377–391.

7. Larsen-Basse, J., – "Binder extrusion in Sliding Wear of WC-Co Alloys". Wear, 1985, Vol. 105, pages 247–256.

8. Pennefather R.C., – "The Solid Particle Erosion of WC-Co Alloys". Msc. Thesis, University of Cape Town, South Africa, 1986.

9. Wentzel E.J *et al.* – "The Erosion–Corrosion Resistance of Tungsten Carbide Hardmetals". International Journal of Refractory and Hard Materials, 1997, Vol. 1 pages 81–87.

10. Brabyn S. *et al.* – "Effects of the Substitution of Nickel for Cobalt in WC Based Hardmetal". 1981, Proceedings of Plansee Seminar, Paper No. HM9.

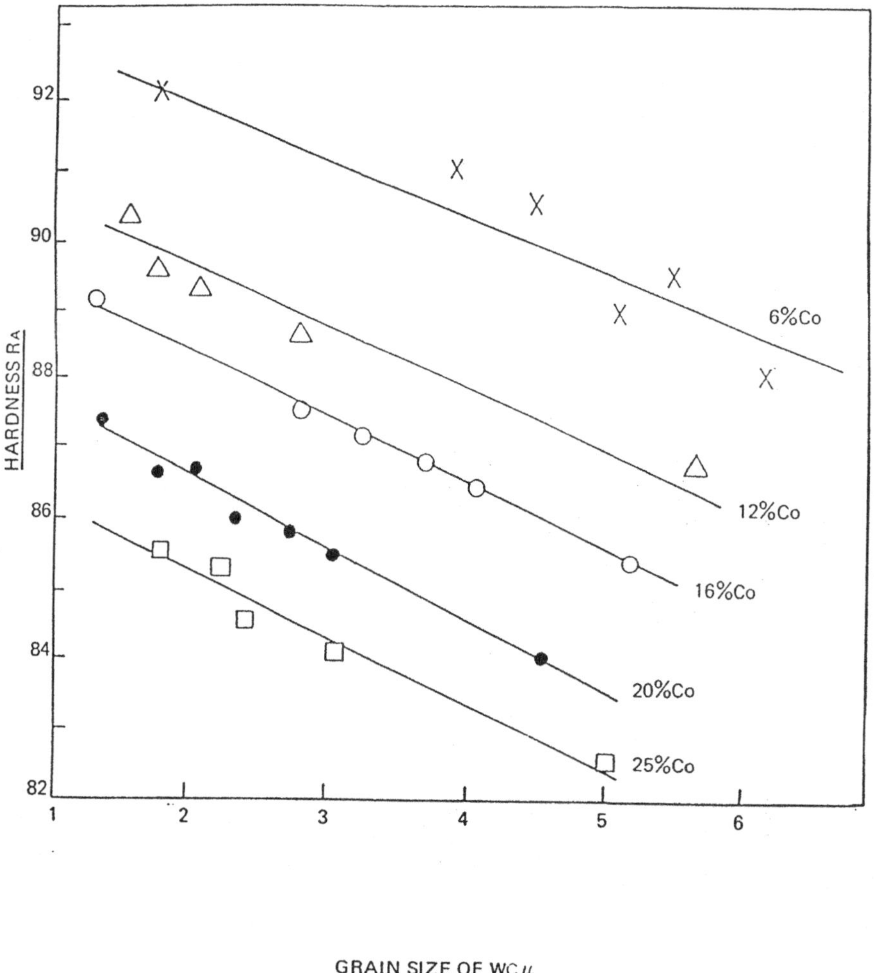

Effect of WC grain size on Rockwell A hardness of WC-Co compositions with different Co contents

Fig. 1

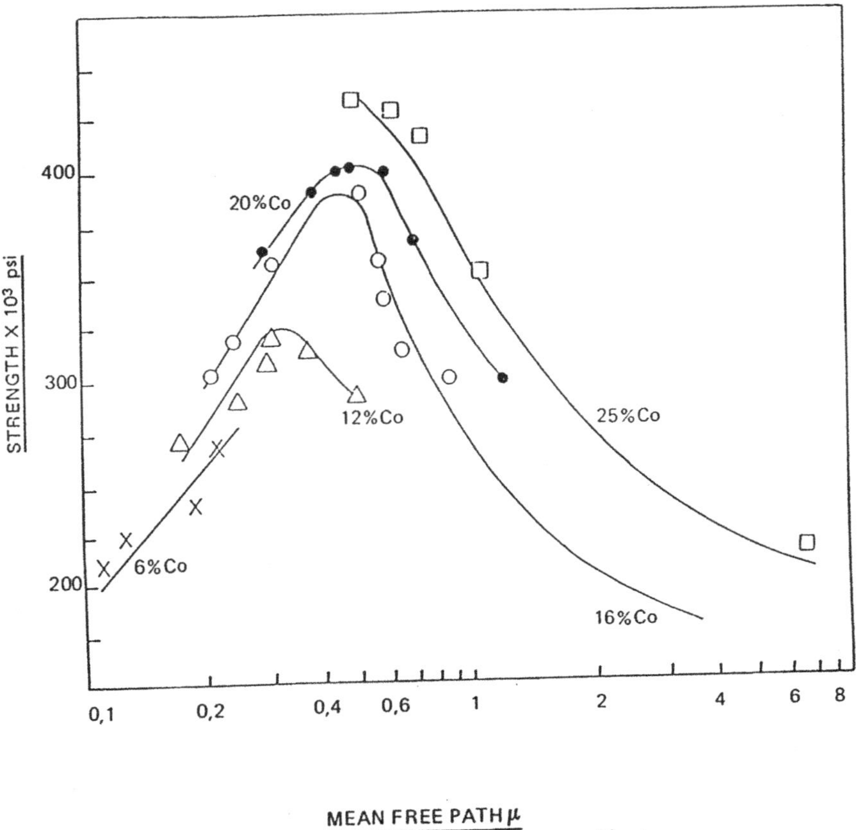

Transverse-rupture strength of WC-Co compositions as a function of Mean Free Path

Fig. 2

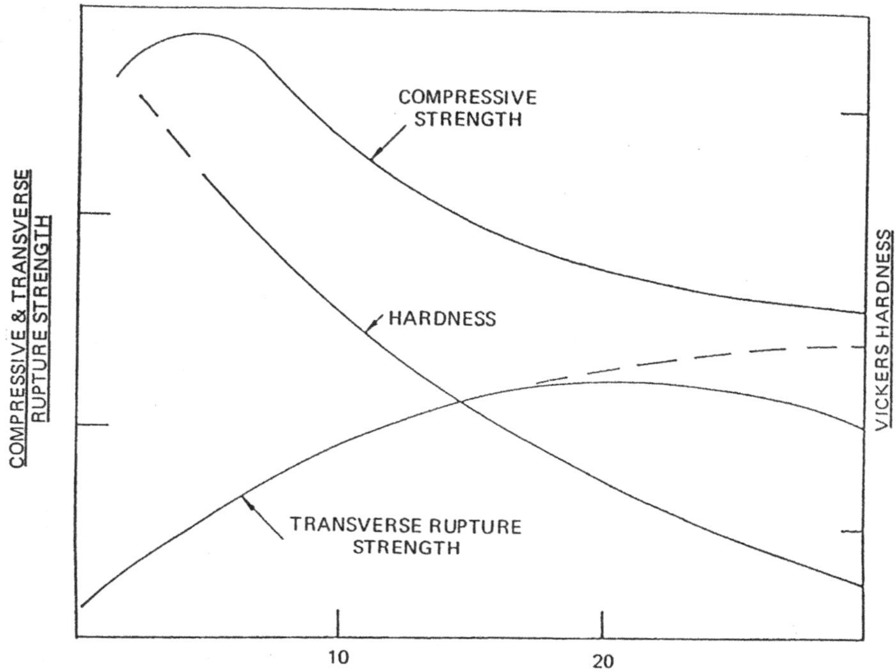

COBALT CONTENT IN wt.%

Effect of cobalt content on the mechanical properties of tungsten carbide-cobalt alloy

Fig. 3

AUTHOR	COBALT MATRIX	WC GRAINS
Blombery et al (1974)	Preferential removal of cobalt (rate controlling step)	This leads to microfracture of the carbide skeleton
Hanson et al (1979)	Preferential removal of binder	
Conrad et al (1982)	Preferential removal of binder shadowing of the cobalt phase by the WC grains at small impingement angles	Cracking of carbides grains
Wright et al (1983)	Preferential removal of cobalt	Ductile erosion of WC grain. Brittle erosion of WC grain by chipping and/or fracture. Pull out of WC grains.
Shetty et al (1984)	Low cobalt content : Cobalt binder removed from triple point boundry junction High cobalt content : Preferential removal of cobalt	Wear rate controlled by transgranular wear of contiguous network of WC grains. The grains are slightly rounded. Grain pull out. Grains show evidence of cracking.
Ball & Paterson (1985)	Ductile removal at high cobalt contents	Brittle fracture of the WC grains at low cobalt contents
Larsen-Basse (1985)	Cobalt extrusion, possibly preceded by cracking of the brittle material	Relaxation of the residual stresses causes the fragmentation of the WC grains

Figure 4

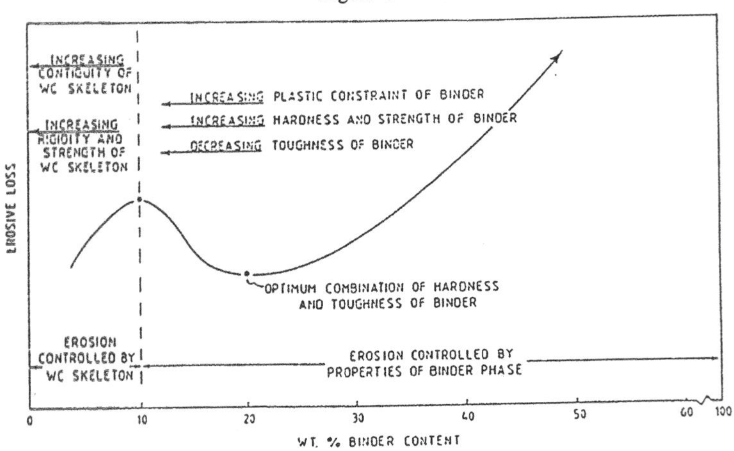

Figure 5

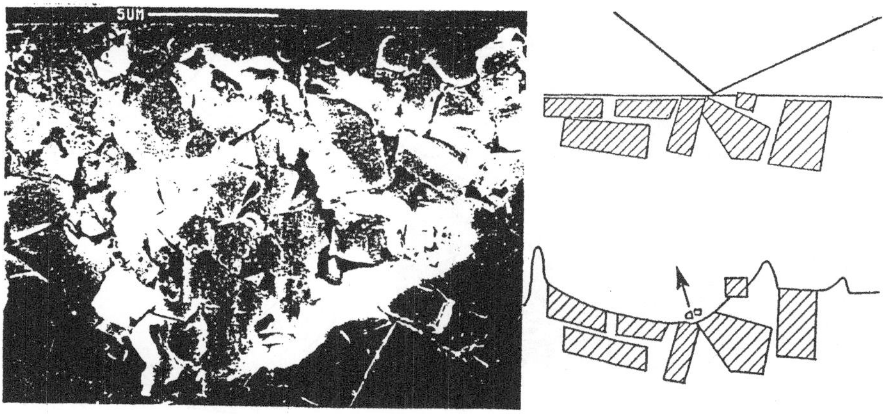

A SEM micrograph of an impact site formed by an 100 μm SiC particle travelling at 40 ms^{-1} and impacting a G1 (7 wt-%, 2.8 μm) specimen at 90°; Schematic representation of the modes considered to occur in material removal

Figure 6

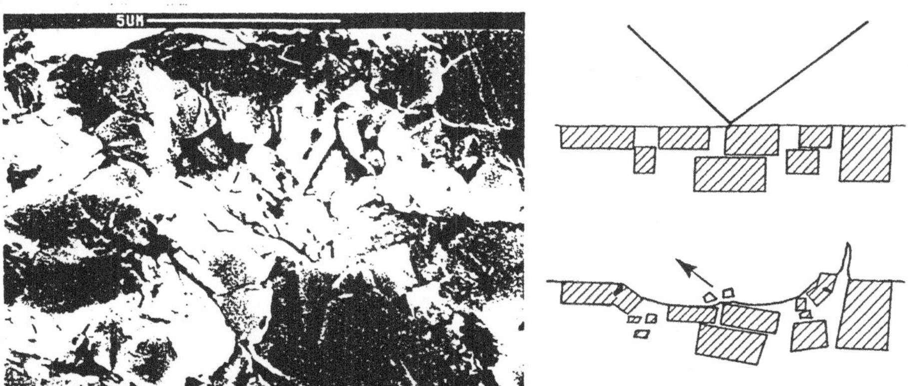

SEM micrograph of an impact site formed be an 100 μm SiC particle travelling at 40 ms^{-1} and impacting a G3 (10 wt-%, 2.82 μm) specimen at 90°; Schematic representation of the modes considered to occur in material removal

Figure 7

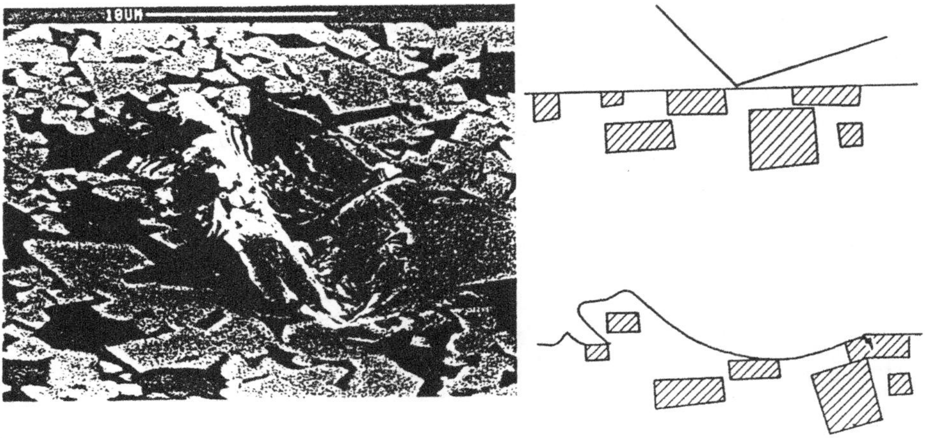

SEM micrograph of an impact site formed by an $100\ \mu m$ SiC particle travelling at $40\ ms^{-1}$ and impacting a I1 (30 wt-%, $2.17\ \mu m$) specimen at 90°; Schematic representation of the modes by which material is considered to be removed.

Figure 8

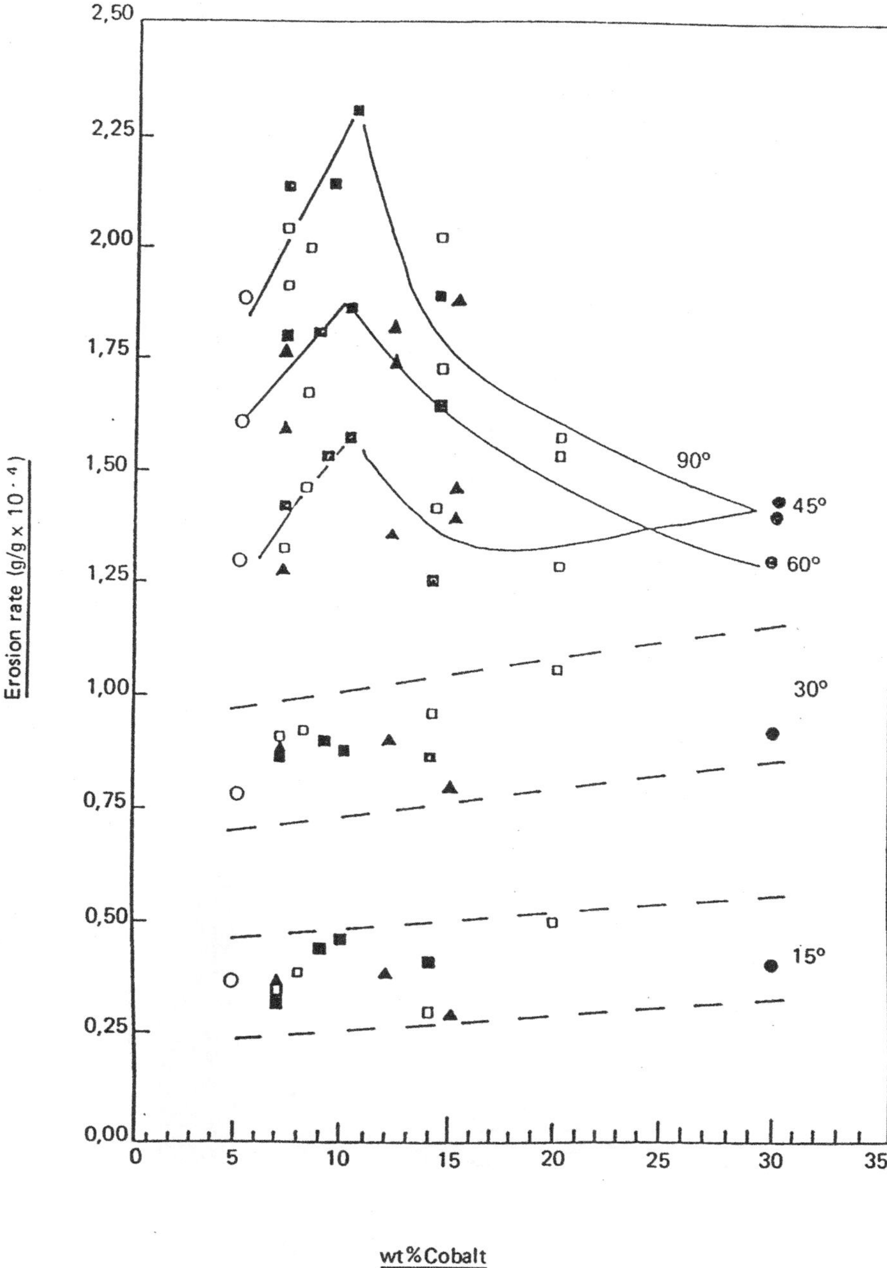

The effect of binder content on erosion rate for the 1,8 μ (open symbols) and the 2,8 μ (Closed symbols) grain sized samples using 100 μ SiC erodant

Fig. 9

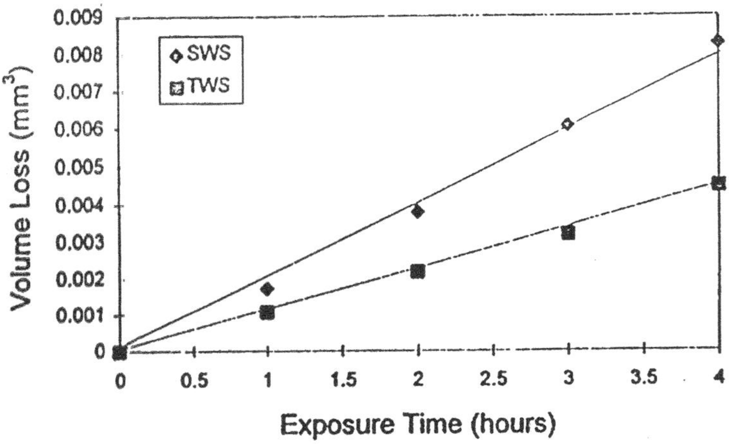

Linear nature of slurry erosion with substitute ocean water and tap water carrier fluids.

Fig. 10

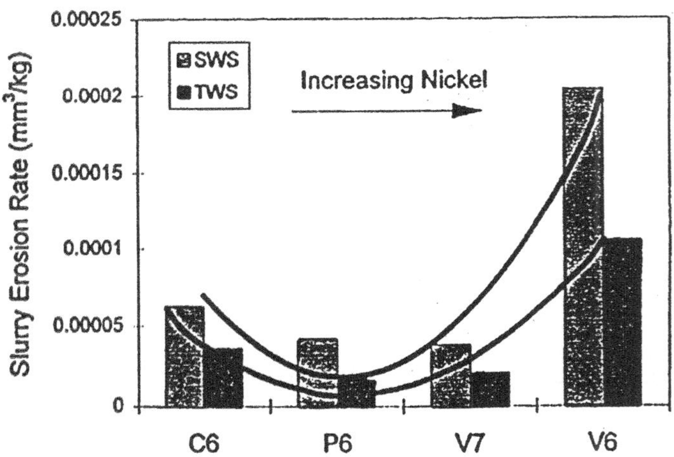

Slurry erosion rates and trends in erosion rate for 6 wt% binder grade cermets in salt water solution (SWS) and tap water solution (TWS).

Fig. 11

PRINCIPALS OF GOOD PRACTICE

Fig 2

ROTATING COMPONENT

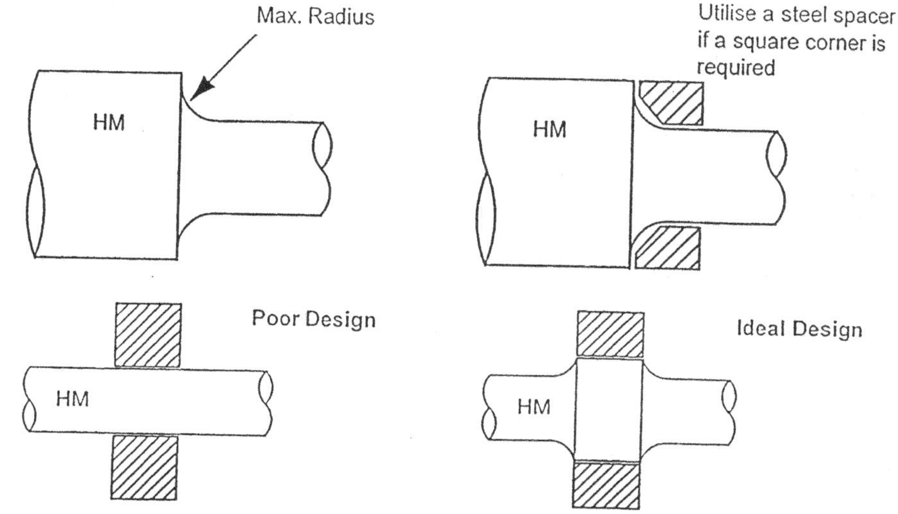

CAPTIVE PLATE

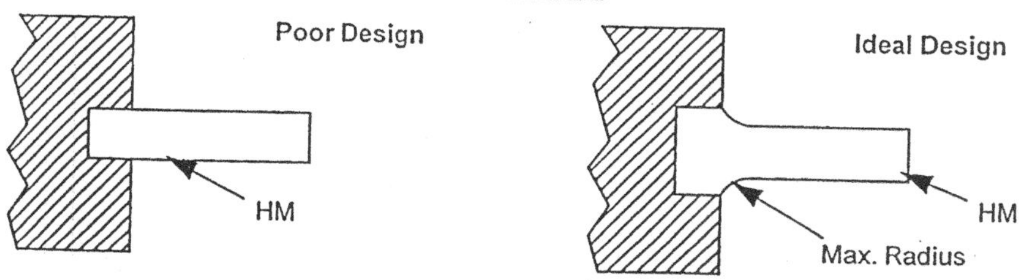

ROTATING DISC

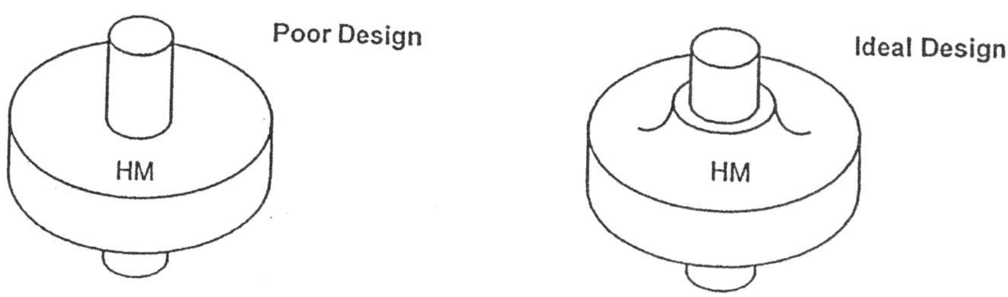

FIG 12

DESIGN CONSIDERATIONS

Fig 1

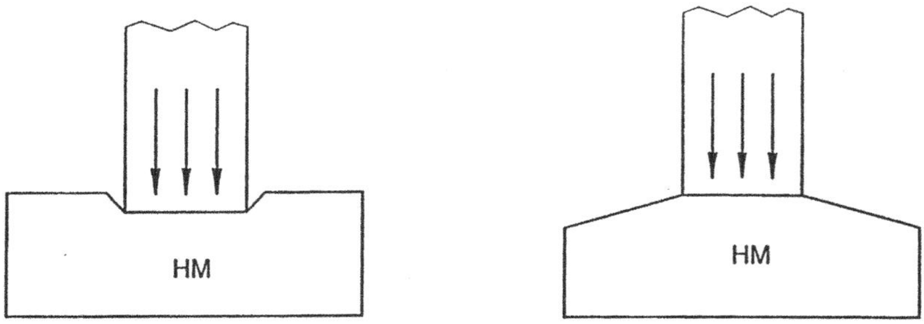

Tensile stress at the load edges are considerably reduced if chamfers are used.

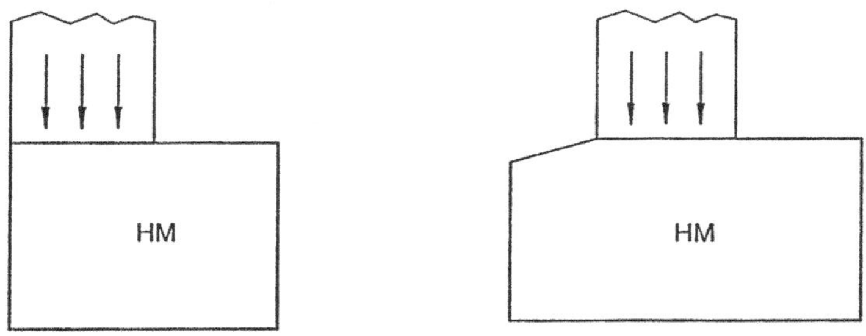

If possible refrain from loading at the edge of a 90 degree corner.

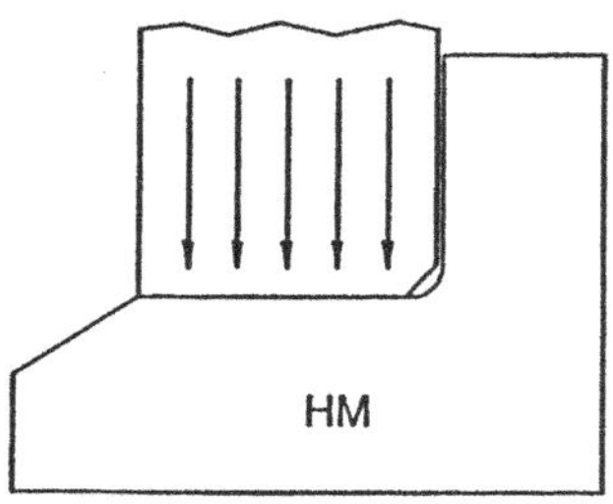

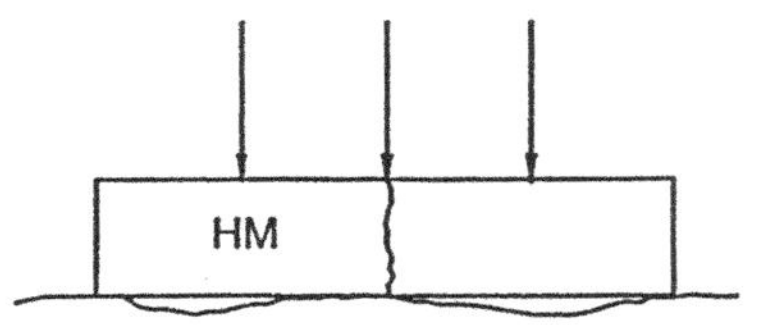

Poor surface contact - Bending stress.

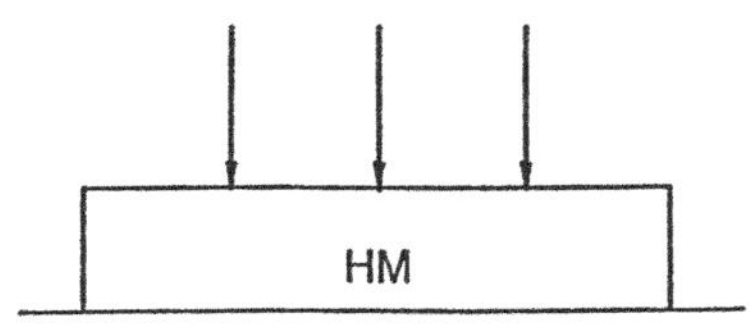

Correct surface contact.

Preferably unmachined fillets are better than ground fillets.

FIG 13

Tungsten Carbide protects pump internals

PUMP ABRASION?
EROSION?.......
DON'T WEAR IT

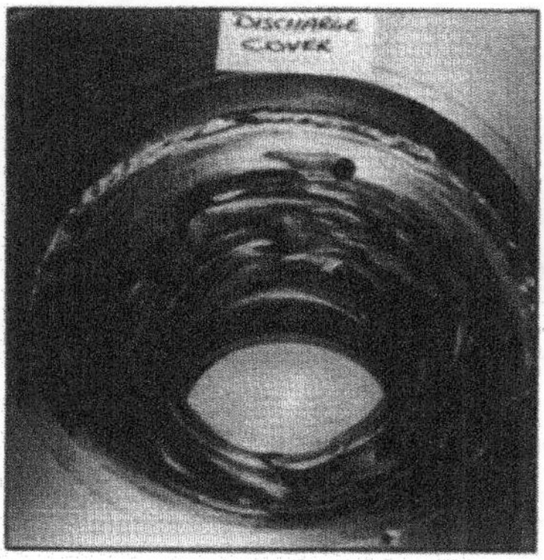

LEFT:
Pump Discharge Cover showing severe abrasion damage.

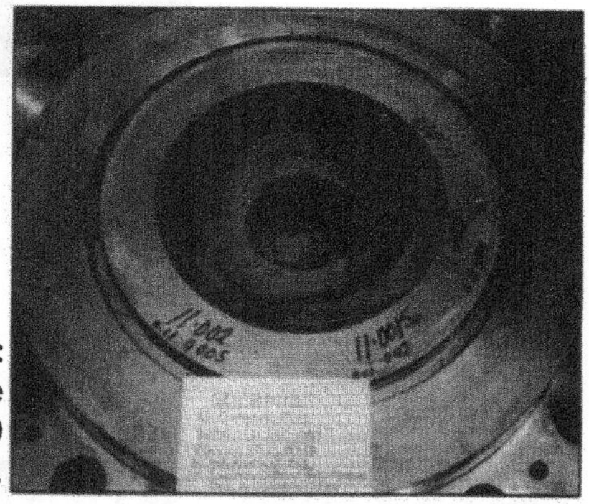

RIGHT:
A Tungsten Carbide Pump Discharge Cover after 9000 running hours, virtually untouched.

One of the greatest enemies of process equipment in the Oil Industry is the presence of sand. Even small amounts of solids in entrained high velocity oil, gas or water streams can quickly cause catastrophic damage to pump internals.

Figure 14.

BOART LONGYEAR

S649/003/99

A case study of Tungsten Carbide in the

PUMP INDUSTRY

(Right), a photograph of a pump without the Tungsten Carbide ring. After just three months, it shows extensive damage to the housing.

The method utilizes a solid tungsten carbide ring which is fitted into the inlet housing of the pump.

The pump which operates at 28 bar pressure delivering gas liquor containing ash, coal and tar, normally has a life of three months.

A pump which was fitted with a tungsten carbide ring has now been running for 22 months and little wear has been experienced in the pump housing. Substantial improvements in the life of the impellor and shaft sleeve have also been obtained.

The improvement in the life of the pump is attributed to the decrease in wear being obtained on the inlet side of the pump due to the fitting of the tungsten carbide ring which in turn reduces wear in other areas of the pump.

(Left), Photograph of a pump fitted with a Tungsten Carbide ring (painted green). This pump shows very little wear after 22 months in operation.

Figure 15

BOART LONGYEAR

Tungsten Carbide

WEAR PLATES

In numerous applications tungsten carbide wear plates provide excellent advantages in prolonging component life over conventional materials. Typical applications include refractory brick moulds, concrete mixers, centrifuges, cyclones, mill liners as well as general metal cleaning and conveying operations.

The large plate in the above photograph is a tungsten carbide plate used in a refractory brick mould which is simply glued into place, thus requiring no expensive or complicated fitting techniques. Other methods of fitting include mechanical fixing, press fitting, brazing, soldering and chemical bonding.

Tungsten carbide hard metals cannot be welded onto wear surfaces. However, in applications where welding is the desired method of attaching the wear part to the surface undergoing wear, various techniques are available for solving the problem. For example, the tungsten carbide plate can be mechanically fitted or brazed onto a steel holder which can then be welded onto the surface, exposing the tungsten carbide to the critical area.

Figure 16.

S649/004/99

Tungsten carbide for abrasion resistant applications

M G GEE and **B ROEBUCK**
Centre for Materials Measurement and Technology, National Physics Laboratory, Teddington, UK

ABSTRACT

Tungsten carbide hardmetals are used in a wide range of abrasion resistant applications. Wear by abrasion occurs through the damage caused by contact with protrusions on bodies that are moving relative to the sample. Often these bodies are particles of abrasive, but they may also be asperties on the opposing surface. In this paper, the variation in abrasion that is observed when key materials and operating parameters such as the composition and grain size of the hardmetal, the compliance of the abrasive media and the hardness of the abrasive, are discussed in the light of the mechanisms of wear that are observed. The use of property maps as a method of choosing and selecting the correct type of hardmetal to obtain the required abrasion resistance is presented.

INTRODUCTION

WC/Co hardmetals are used in many applications where resistance to abrasion is important, combining high hardness with reasonable levels of toughness as a consequence of their microstructure. They are composite materials which have WC grains embedded in a metallic binder which is normally Co, but can also be Ni or Ni/Co alloys. A powder metallurgy liquid phase sintering process is used for their production.

By changing the main microstructural parameters of binder volume content (0 to 25%) and WC grain size (0.5 to 5 μm) a range in hardness values from 800 to over 2000 HV30, and Palmqvist toughness from 7 to over 40 MN $m^{-1.5}$ can be achieved.

Because of the industrial importance of these materials, there has been interest over a number of years in the development of techniques that study their abrasion resistance in the laboratory [1-10]. This has resulted in the development of abrasion tests such as the

ISO 12962/ASTM B611 test which is described as suitable for hardmetals used in mining applications [4,10,11], and the ASTM G65 dry sand abrasion test.

In many of the applications where resistance to abrasion is important, the contact loads between the abrasive and the component are high. When this occurs, fracture of the abrasive often takes place during the abrasion process.

By contrast in low stress abrasion the loads applied during abrasion are much lower, and fracture of the abrasive does not occur.

This paper gives the results of tests performed with a high stress (ASTM B611) and low stress (ASTM G65) abrasion tests. The variation in abrasion that is observed when key materials and operating parameters such as the composition and grain size of the hardmetal, the compliance of the abrasive media and the hardness of the abrasive is discussed in the light of the mechanisms of wear that are observed. The use of property maps as a method for choosing and selecting the correct type of hardmetal to obtain the required abrasion resistance is presented.

ABRASION

Abrasion is defined by ASTM G40 as "wear due to hard particles or hard protuberances forced against and moving along a solid surface". It is distinguished by the normally low velocity of the contacting materials, and unlike erosion cannot occur without the contact between two solid bodies.

The magnitude of abrasion depends on:

- the abrasive material, hardness, size and shape (rounded or sharp corners);
- the constraint on movement of the abrasive (fixed or free to rotate);
- the compliance of the counterbody;
- the load;
- the relative speed of the abrasive and the sample;
- the abrasive concentration;
- the test environment.

The effect of the compliance of the counterbody is particularly important, Figure 1, where it can be seen that with a high compliance counterbody there is load sharing and low contact stresses in comparison with a low compliance counterbody where there is little or no load sharing and high contact stresses.

ABRASION TESTS

Two different test systems were used. These were the ASTM B611 system (Figure 2) which uses a steel wheel as a counterbody and aluminium oxide as the abrasive. In this test the abrasive is picked up by the steel wheel from the wet slurry and is carried around abrading the test sample.

In the modified ASTM G65 test which was used in these experiments is shown in Figure 3. In this test the abrasive is fed onto the test wheel by a slotted drum and feed chute, which can be rubber rimmed or steel, just before the test sample. Fluid can be fed onto the wheel just before the abrasive to carry out tests in a wet condition.

Full details of the test conditions are given in Table 1.

Tests were performed on a range of WC/Co hardmetals with a range of hardness from 800 HV30 to 2160 HV30. These had a range of different grain sizes and cobalt binder phases.

Table 1 - Test Conditions

	ASTM B611	**Modified ASTM G65**
Compliance of Counterbody	Low	Low or high
Abrasive	600 μm alumina and silica	200 μm silica
Test Load	200 N	130 N
Test Duration	1000 revolutions	8000 revolutions
Wheel	165 mm diameter steel wheel	200 mm diameter rubber rimmed or steel wheel
Environment	De-ionised water and simulated sea water	Dry or de-ionised water

In both tests the wear volume was calculated from the mass change on wear and the density of the materials, and the worn surfaces were examined by optical and scanning electron microscopy. In the case of the ASTM B611 tests, samples were sectioned perpendicular to the wear surface to determine the mechanisms of wear that had occurred.

TEST RESULTS

ASTM B611

Results of the ASTM B611 steel wheel abrasion test are shown in Figure 4. For both abrasives, there was a strong correlation of wear volume with the hardness of the hardmetal, with a reduction in wear volume of over two orders of magnitude for the range of materials that have been examined. The spread in results about the trend line is remarkably small. The results can be described quite well by the relationship

$$V = Ae^{-BH}$$

where V is the wear volume, H is the hardness of the hardmetal and A and B are constants. At low values of hardmetal hardness, the experimental results are slightly below the values predicted by the extrapolated relationship.

The volume of wear for the silica are about an order of magnitude lower than those for the alumina abrasive.

It is worth emphasising that the very tight sets of results was obtained for a diverse set of hardmetals with different volume fractions of Co binder phase and different grain sizes.

No effect of using simulated sea water instead of distilled water was seen (Figure 5).

Modified ASTM G65
Figure 6 shows the results of the rotating wheel tests performed on hardmetals. There was a decreasing trend in wear with increasing hardness in all three conditions, and the wear volume always increased in the order wet rubber → dry rubber → wet steel. It is noticeable that the slope of the curves decreased as the severity of the wear decreased, and that the scatter in results increased as the wear decreased. This latter point may be due simply to the reduction in magnitude in wear, with a consequent increase in the effect of limited measurement accuracy at low wear.

An exponential dependence of wear volume on hardness of the form described earlier for the ASTM B611 tests was always found.

Comparison between two tests
The results of the steel wheel modified ASTM G65 test are compared with the ASTM B611 results in Figure 7. Although the abrasives are different sizes, and the test loads are different, it can be seen that the trends in the results are almost identical.

OBSERVATION OF WORN SURFACES

Sample Appearance
A typical wear scar from the ASTM B611 test is shown in Figure 8a. It can be seen that the scar shape is conformal to the shape of the steel wheel. The leading edge of the wear scar is noticeably sharper than the trailing edge, which has a feathery appearance.

Figure 8b shows a typical wear scar produced on a hardmetal sample with a wet rubber wheel. The wear scar was quite shallow and relatively smooth. Wear scars for tests performed with the dry rubber wheel were quite similar, but had a duller appearance.

ASTM B611 Wear Surface
Wear Surface
Wear surfaces were always dull and rough. Examination at high magnification showed a number of features common to all the wear surfaces.

Thus there was always considerable fracture damage to the WC grains (Figure 9a). Cracks in individual grains of WC was often aligned with the perceived crystallographic axes of the crystal. In many cases crushing of the WC grains had taken place with multiple cracks breaking up what had previously clearly been single grains into multiple fragments of WC. Fracture to the WC grains became less obvious as the grain size of the hardmetal decreased, but this may simply be due to the increasing difficulty in distinguishing evidence for fracture as the grain size becomes very small.

Many of the small fragments of WC had been re-embedded into binder phase regions of the worn surface; the size of some of these WC fragments was quite small at about 20-50 nm (Figure 9b).

In some places fragments of material had clearly broken away from the wear surface (Figure 9c). The depressions that remained were larger in size than a single grain, indicating that clumps of grains had broken away from the surface.

Cross Sectional Analysis of ASTM B611 Samples
Some cross-sections were made by using a diamond saw to cut samples in half perpendicular to the wear surface. The sections were polished carefully to preserve the geometry of the corner edge between the wear surface and the polished surface. The samples was then mounted in the SEM such that the wear surface and the polished face gave equal intensity in image formation from the secondary electron detectors.

The result was a series of images where the left hand side of the image corresponds to the edge of the worn surface, and the right hand side was the polished cross-section through the sample. (The disadvantage of the adjustment of the sample position to equalise image intensity between the worn surface and the polished section is that it makes it difficult to determine where the corner is because of the loss of depth cues in the image.)

The examination of cross-sections showed the following features.

- The worn surface was sometimes deeply scored, this is very evident in the cross-section, but can be difficult to detect in conventional SEM imaging of the wear surfaces due to the high depth of focus of these instruments.

- The area where fragments of WC were re-embedded into the binder phase could be quite deep on the scale of the microstructure (depth of several grain dimensions) (Figure 10). However, the thickness of these areas was not homogeneous across the whole surface of the sample.

- Some grains at the surface were rounded.

- Sub-surface cracks parallel to the wear surface were frequently observed at a depth of a few grains beneath the wear surface (Figure 10). Again these cracks were not always observed, and were not continuous on samples where they were seen.

Although the regions where fragments of WC had re-embedded into the binder phase and the sub-surface cracks were observed in many samples, the number of observations was too small to attempt to relate the details of their position and size to features of the microstructure such as grain size and binder volume fraction.

ASTM G65 Wear Surface
Figure 11 show higher magnification micrographs of hardmetal samples tested with wet rubber, dry rubber and wet steel test wheels. There was an increase in the degree of fracture and damage to WC grains as the WC grain size of the hardmetal increased, and an increase in the degree of damage as the severity of wear increased with changes in test condition from wet rubber wheel $\rightarrow$ dry rubber wheel $\rightarrow$ steel wheel.

DISCUSSION OF TEST RESULTS

In the steel wheel abrasion tests, deep scratching to the samples takes place by the action of the highly loaded abrasive particles. This causes plastic deformation to both the WC and to the binder phase, and also causes considerable fracture and fragmentation to the WC grains. Although some of these small fragments of WC are removed from the wear surface, many remain in the materials structure. There are two possible ways that this can occur; either the fragments are re-embedded in the structure as further contacts occur, or the binder phase is pushed out or extruded through the network of cracks that is formed in the WC grains [5,6]. No direct evidence for extrusion was observed in this work, and the examination of sub-surface sections did not show any obvious sign of loss of binder phase from the surface areas.

Perhaps more significantly, cracks are generated and propagated a few grains beneath the surface of the wear surface of the hardmetals. These cracks were seen in all the samples that were sectioned, but were not continuous across the complete width of the sample. Although the possibility of the generation of sub-surface cracks through indentation mechanisms such as Palmqvist cracking has been suggested [8], the form of these cracks suggests that they are more likely to be formed by sub-surface fatigue mechanisms. These cracks weaken the structure of the material such that further fragments of the samples are removed by future abrasion events.

WC grains in samples which have been scratch tested show similar evidence for fracture and deformation damage to WC grains in the steel wheel abrasion test. However, because only a single abrasion events occur, the surface is much less damaged and damage to individual grains can be seen more clearly [12,13]. It was also found that there was a correlation between the size scratches and conventional indentation hardness of the hardmetals.

It is considered that the rate of wear is controlled by the processes that occur during the formation of scratch grooves during individual abrasion events. Material is actually removed by several processes which scale with the process of groove formation including the formation of sub-surface cracks by a fatigue process under repeated abrasion, and the direct fracture and fragmentation of WC grains.

In the modified ASTM G65 tests, the degree of fracture decreased as the compliance of the test wheel was increased. This is because the steel wheel produced more severe abrasion damage and fracture. This is likely to be because the individual abrasive particles in contact with the testpiece are more highly loaded than those on a compliant rubber surface. There was a reduction in wear for tests carried out wet compared to tests carried out dry. The reason for this is not completely clear. Factors which may contribute are the cooling of the abrasive contacts by the water and consequent reduction in temperature at the wear surface, the lubricating effect of the water, and possible changes in the dynamics of motion of the abrasive particles in the presence of water compared to air. However, although the cooling effect may contribute, it is not felt to be too important, as the temperature of the wheel did not rise that sharply during tests.

PROPERTY MAPPING

Property maps combine a knowledge of the structure with the basic mechanical properties of the hardmetal. This provides a tool which can be used to allow for the effects of differences in structure and properties to be easily compared and provides a benchmark for the evaluation of new materials.

The approach that is used is to combine the knowledge of the materials behaviour with empirical results, in the case of this paper, of abrasion tests.

Two types of property map are presented.

In the first type the volume of wear is plotted against a microstructural feature. In this case two microstructural features were evaluated, WC grain size, d, and wt% cobalt content, W. It is recognised that other parameters are important, such as the composition of the cobalt binder phase or the size distribution of WC grains. However, for this report the property maps were developed assuming a composition approximately in the centre of the two-phase WC/Co region and assuming a conventional WC grain size distribution with the arithmetic number mean intercept, d, as the parameter characterising WC grain size.

In the second type of property map the wear volume was compared against the hardness and toughness to give two maps:

$$A \ \text{v} \ H \qquad A \ \text{v} \ T$$

For property mapping the Hall-Petch correlation between H and d was used to derive the expression relating Abrasion resistance (volume loss), A, to d, since

$$\log A = g + hH$$

where g and h are constants, A is in cm^3 and H is the HV30 value.

$$\text{but} \qquad H = e + fd^{-0.5}$$

$$\text{thus} \qquad \log A = i + kd^{-0.5} \qquad (1)$$

where i and k are constants, and d is in µm.

The abrasion resistance measurements for the CAM3/4/9 materials are shown plotted against grain size in Figure 12 and inverse square root of the grain size in Figure 13 with the abrasion resistance (volume loss) plotted on a logarithmic scale. Analysis of the data in Figure 4 allowed an expression to be written for the variation of abrasion resistance, A, with grain size, d, for 6 < Co < 25 wt% as follows:

$$\log_{10} A = (0.19 - 0.016 * \%\,\text{Co} - \frac{(2.075 - 0.133 * \%\text{Co} + 0.0022 * (\%\,\text{Co})^2}{\sqrt{d}} \qquad (2)$$

Expression (2) can be plotted as a property map, Figures 14 and 15 (with x as d and x as $d^{-0.5}$ respectively).

A property map using Co wt% content as the variable can also be plotted from expression (6): This is shown in Fig 16 for different values of the WC linear intercept grain size.

The comparative map, Fig 17, shows the relation between Abrasion (vol loss) and hardness. The form of this map is similar to that in Fig 4 showing the measured data but there is more scatter than the experimental results. More work is probably required to derive accurate values for the coefficients in the expression relation Abrasion (vol loss) to grain size, particularly for the coarse grain size high Co materials.

CONCLUSIONS

The abrasion resistance of hardmetals has been studied using ASTM B611 and ASTM G65 tests. A systematic variation of abrasion volume has been found with hardmetal hardness, and in the ASTM G65 tests with the compliance of the test systems.

The concept of property mapping has been presented with a preliminary derivation of an abrasion property map for hardmetals.

Abrasive wear in hardmetals is predominately caused by fracture processes. These include:
- Fracture of individual WC grains
- Delamination
- Removal of clusters of WC grains

However, other processes also take place such as:
- Re-embedding of WC fragments
- Removal of binder phase from the surface layers
- Polishing.

ACKNOWLEDGEMENTS

The authors would like to acknowledge the support of DTI through the Materials Measurement Programme, Characterisation of Advanced Materials.

REFERENCES

1. E A Almond, L A Lay, M G Gee. Comparison of Sliding and Abrasive Wear Mechanisms for Cemented Carbides and Ceramics, Proceedings of 2nd International Conference on the Science of Hard Materials, Institute of Physics Conference Series No 75: p919, Adam Hilger, 1986.

2. J Larsen-Basse and N Devnani, Science of Hard Materials, Proceedings of 2nd International Conference on the Science of Hard Materials, Institute of Physics Conference Series Number 75, pp 883-895, 1986.

3. J Larsen-Basse, C M Shishido and P A Tanouye, The Journal of the Australian Institue of Metals, 19(1974)271-275.

4. H Feld and P Walter, Powder Metallurgy International, 7(1975)188-190.

5. J Larsen-Basse, International Conference on Wear of Materials, ASME, (1985), 39-44.

6. J Larsen-Basse, Materials Science and Engineering, 13(1974)83-91.

7. K Jia and T E Fischer, Wear, 200(1996)206-214.

8. S F Wayne, J G Baldoni and S T Buljan, Tribology Transactions, 33(1990)611-617.

9. J Larsen-Basse, Powder Metallurgy, 16(1973)1-32.

10. D G F Quigley, S Luyckz and M N James, Int J Refractory & Hard Materials, Vol 15, pp73-79, 1997.

11. ASTM B611-85, Standard test method for the abrasive wear resistance of cemented carbides, Annual Book of ASTM Standards, 02.05(1995)326-327.

12. M G Gee, A Gant and L Van der Berg, Scratch Testing and Fine Scale Abrasion of Hardmetals and Ceramics, NPL Report CMMT(A)175, May 1999.

13. M G Gee and L Van der Berg, Low Load Multiple Scratch Testing of Hardmetals and Ceramics, CMMT Measurement Note MN 45, May 1999.

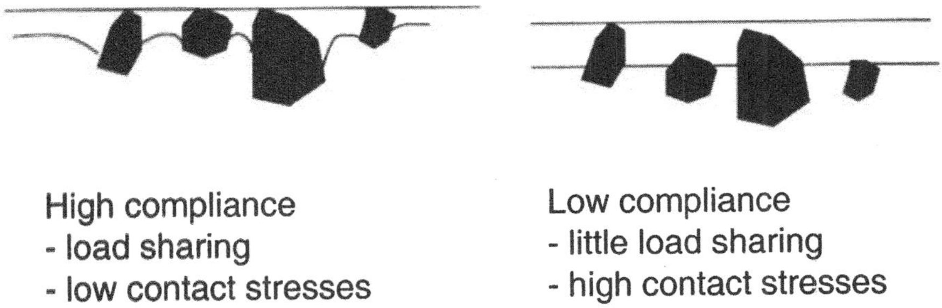

Figure 1 Load sharing with different counterbody compliance.

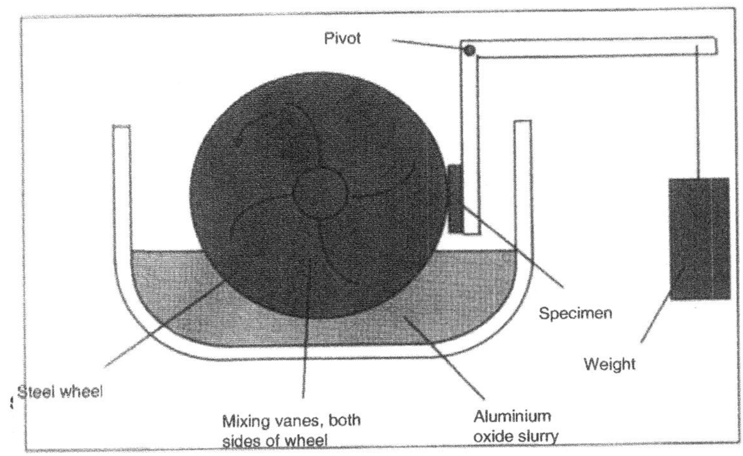

Figure 2 ASTM B611 test system.

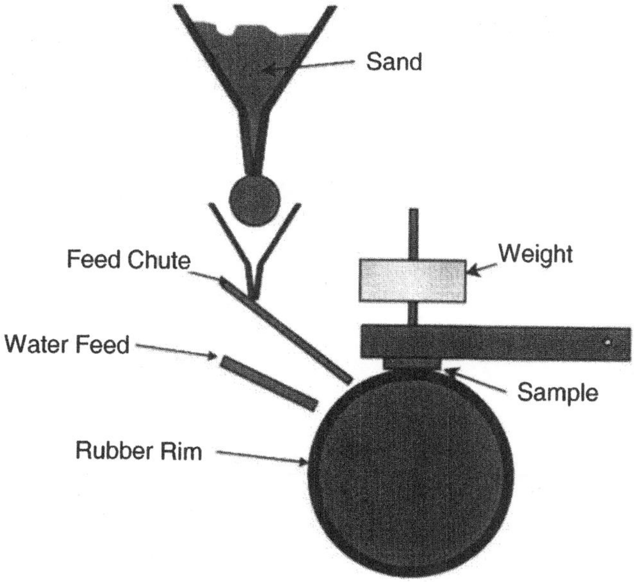

Figure 3 Modified ASTM G65 test system.

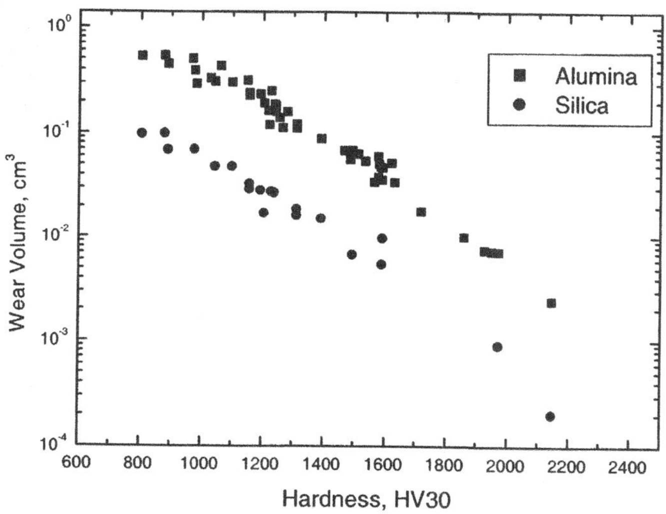

Figure 4 Results of ASTM B611 tests for alumina and silica abrasives.

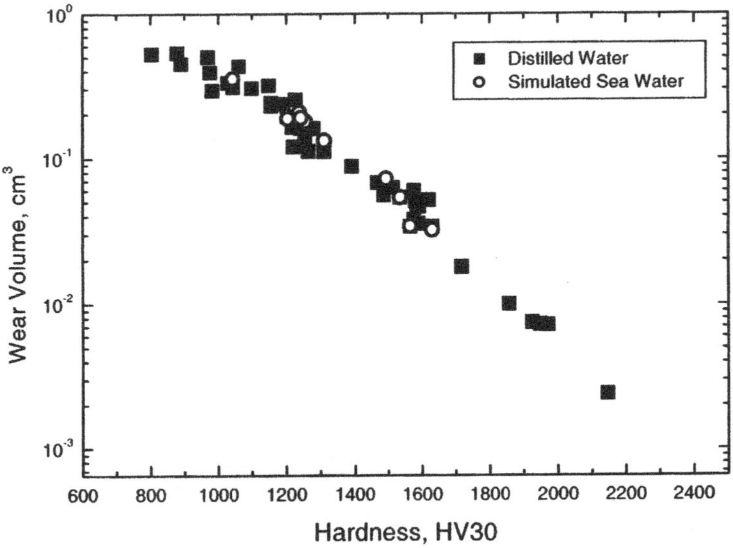

Figure 5 **Effect of sea-water on ASTM B611 test results.**

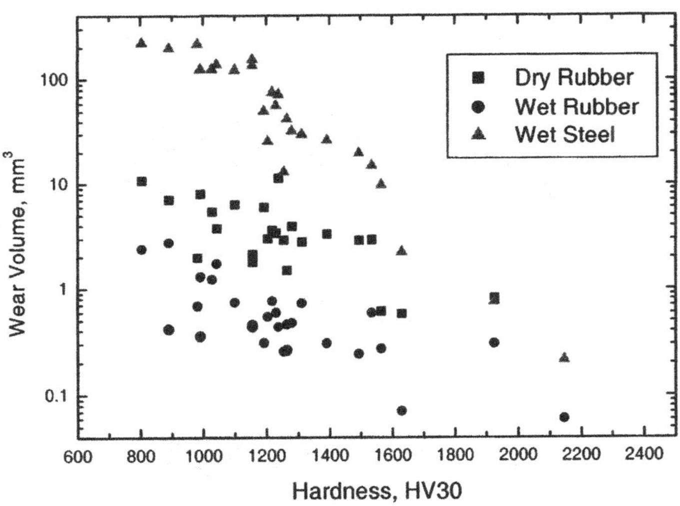

Figure 6 **Results of modified ASTM G65 tests.**

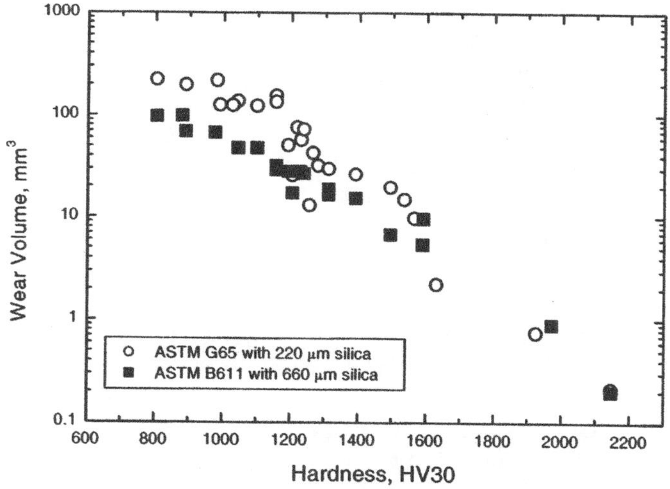

Figure 7 **Comparison between ASTM B611 and ASTM G65 test results.**

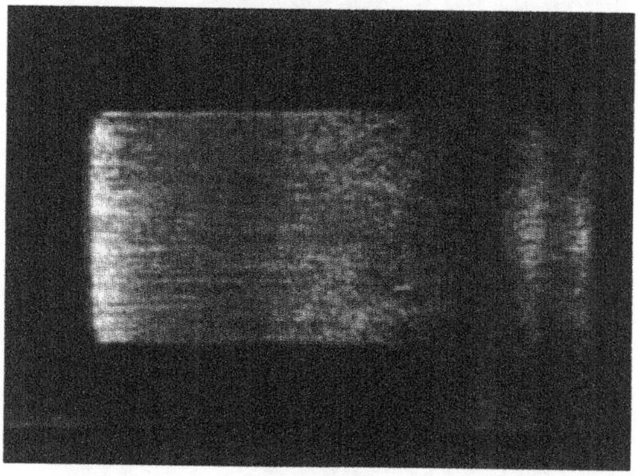

a)

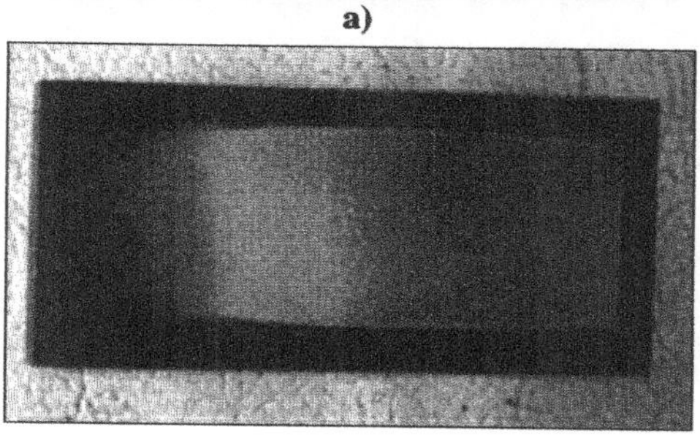

b)

Figure 8 **Typical wear scars, (a) ASTM B611 sample, (b) Modified ASTM G65 sample.**

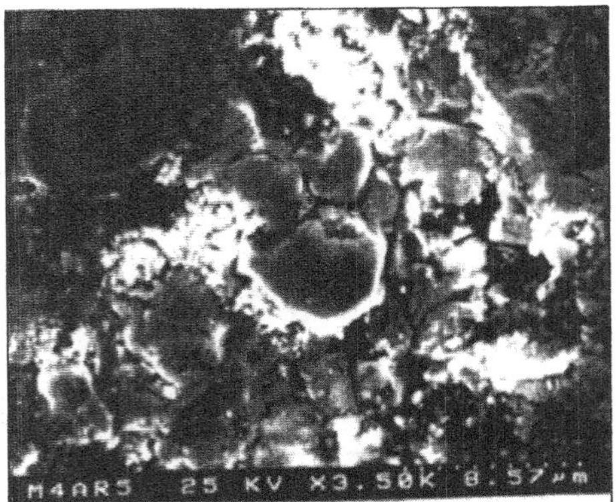

a)

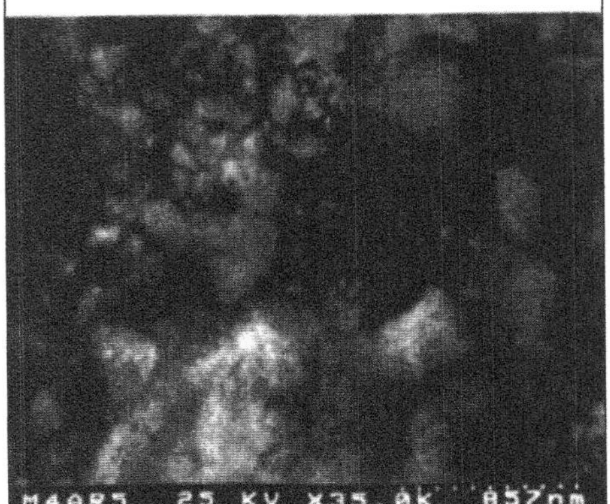

b)

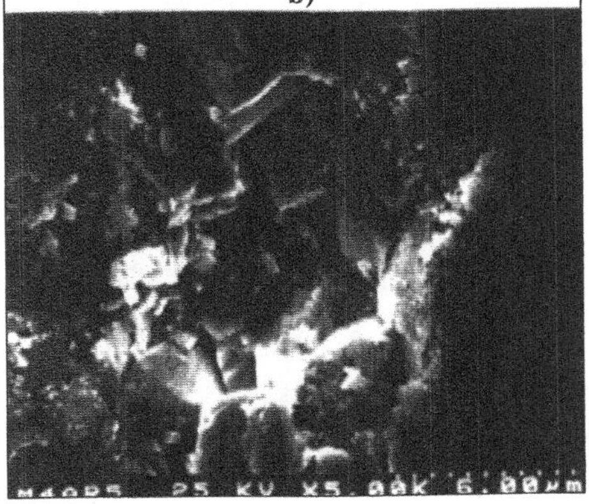

c)

Figure 9 Typical wear surfaces of ASTM B611 sample
(a) fractured grains, (b) re-embedded WC fragments, (c) fracture of fragment.

Figure 10 **Delamination revealed by cross-section through ASTM B611 test sample.**

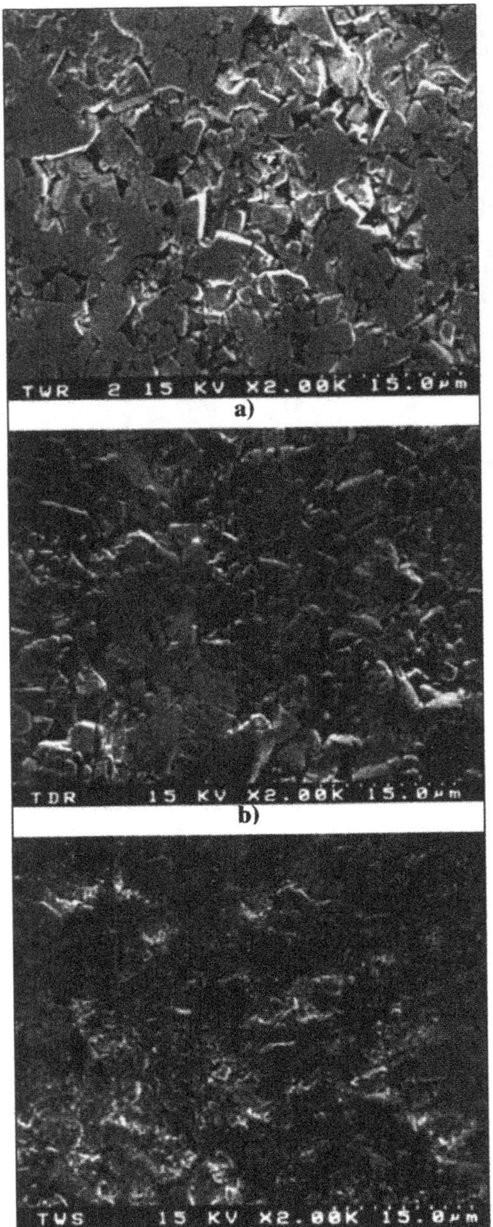

c)

Figure 11 Wear surface of modified ASTM G65 test samples (a) wet rubber, (b) dry rubber, (c) wet steel.

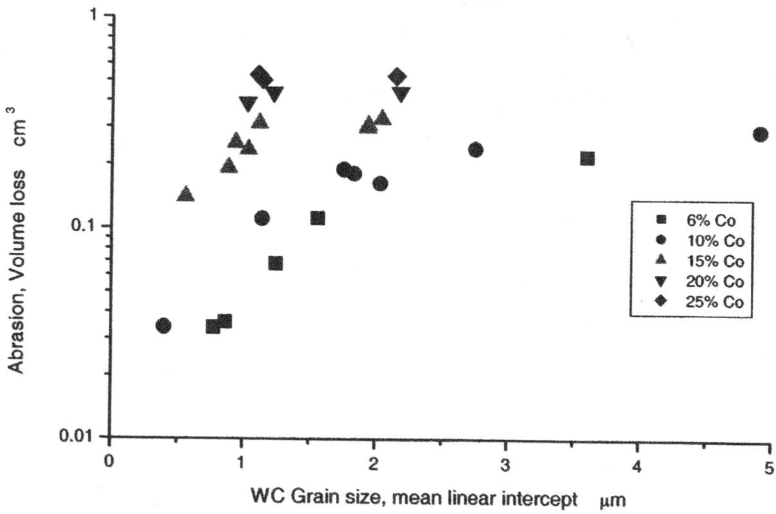

Figure 12 Empirical abrasion wear volume - grain size property map.

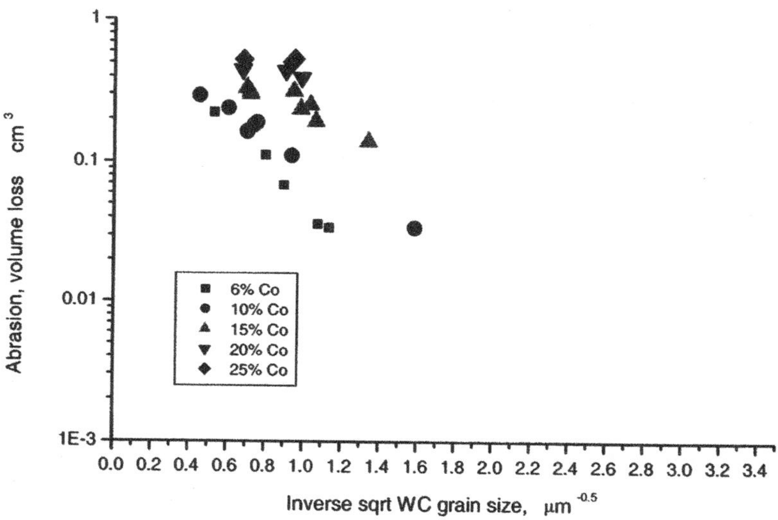

Figure 13 Empirical abrasion wear volume - inverse square root grain size property maps.

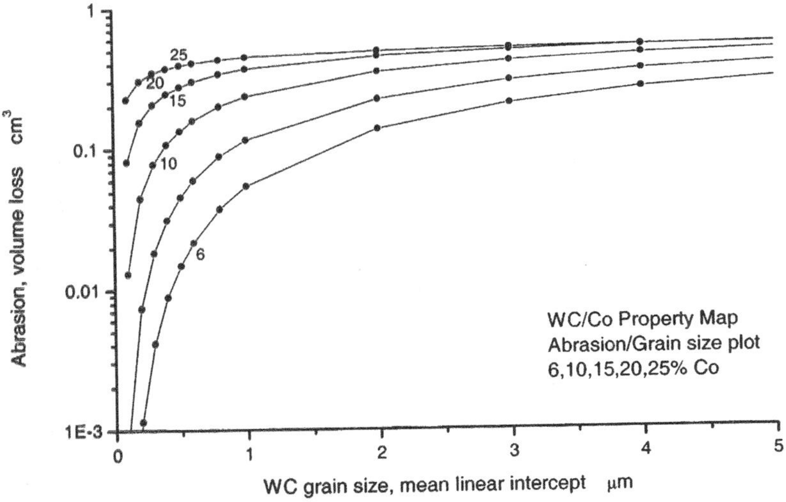

Figure 14 Fitted abrasion wear volume - grain size property map.

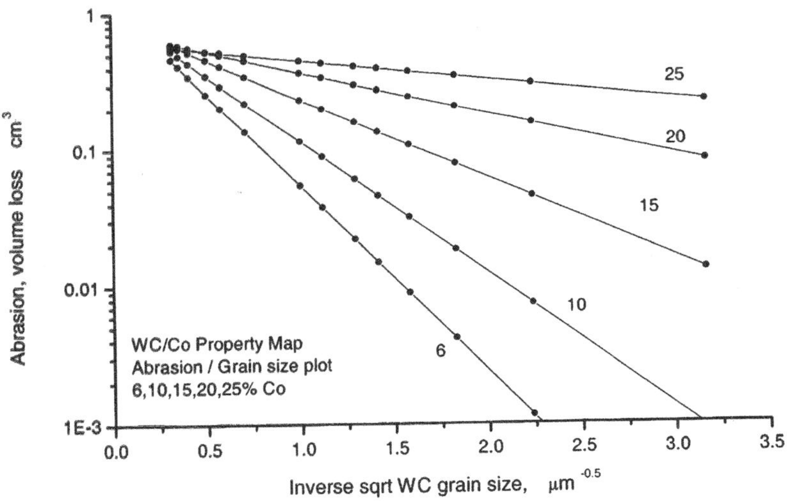

Figure 15 Fitted abrasion wear volume - inverse square root grain size property map.

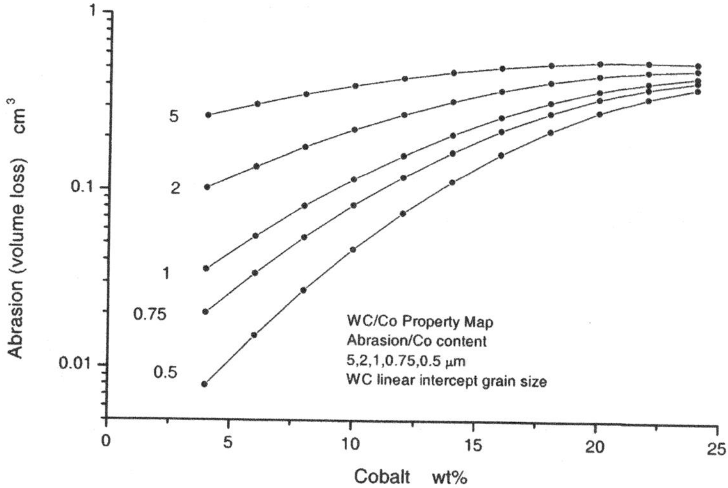

Figure 16 Fitted abrasion wear volume - cobalt wt% property map.

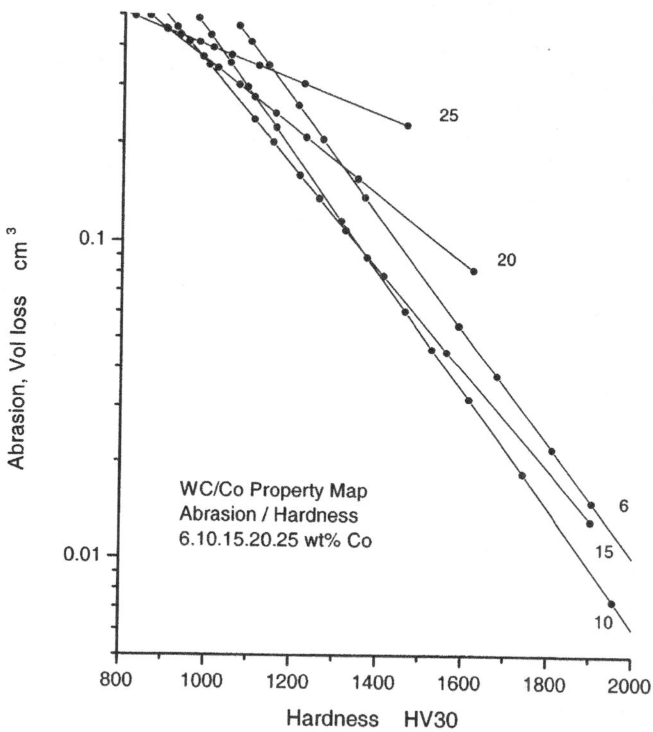

Figure 17 Comparative abrasion - hardness property map.

S649/005/99

Designing with powder metallurgy (PM) for increased fatigue life

S BENGTSSON
Höganäs AB, Sweden

SYNOPSIS

Factors influencing the fatigue life of PM components and guidelines for design are reviewed. The different stages in the fatigue process are described, such as the crack initiation, crack propagation and fracture. The influence of porosity, powder/alloy system, notches and density is discussed as well as influences of secondary operations including heat treatments, and surface densification.

INTRODUCTION

The design process must always start with an assessment of the function of the component. In most cases the function is geometrical, i.e. a connecting rod is supposed to keep the distance between the piston pin and the crankshaft constant. In order to do so the component will be subjected to various loads. It can in many cases be very difficult to estimate the amplitude and direction of these loads. However, if the required safety factor should not be exceedingly large the service loads must known to a sufficient degree of accuracy.

The fatigue properties are traditionally described in the S-N (Whöler) diagram. This diagram is constructed by making a number of tests at different stress amplitudes and constant mean stress. The number of cycles to failure is a measure of the resistance to fatigue failure. For most materials there is a lower limit where no fatigue failure occurs even if the testing goes on for a very long time. This level is called the fatigue limit or endurance limit. Since the failure is a statistical process the fatigue limit is usually described as the stress amplitude where 50% of the test bars survive. When a component is designed a measure of the spread of the fatigue limit must be available in order to facilitate calculations of a suitable safety margin. In this text the 50% limit will be used, but other limiting probabilities exists (e.g. 99%). Usually fatigue data does not exist for the type of load the component is subjected to in terms of stress amplitude and mean stress. However, if some fatigue data exists, a Haigh diagram can be constructed. In this diagram the relation between stress amplitude and mean stress can be estimated, see Figure 2.

The fatigue process can be described as a process where the weakest part of the material fails. In most cases the weakest point is a material defect or artefact somewhere in the component. If the component is large the probability of finding such a defect increases. This means also that there is a difference between axial loading where the entire cross section experiences full load and bending where only the outermost part of the cross section experiences the full stress, see Figure 1.

Notches acts as stress concentrations, but the stressed volume is small compared to the cross section, which means that the reduction in fatigue limit is not as severe as the stress concentration indicates. For a Fe-2Cu-2,5Ni P/M steel the fatigue limit decreased by a factor of 1,6 for a stress concentration of 2,8. In this case the notch sensitivity was 0,6. Unfortunately, this can not be taken as a general truth, but must be evaluated for each material and density level. The notch sensitivity increases with the density. The fatigue limit for P/M steels is usually of the same order as nodular cast iron and have approximately half the fatigue limit of wrought steels. However, if the notch factor (k_t) is above 2,5 the P/M materials have a fatigue limit in the same range as wrought steels.

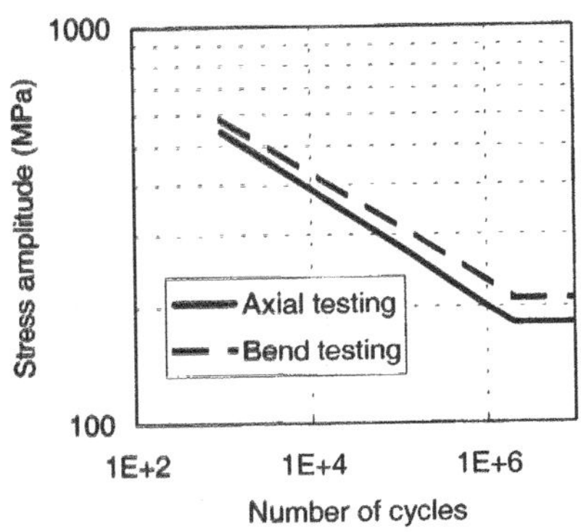

Figure 1. S–N diagram, showing difference between axial- and bend fatigue test. Material: Distaloy AE +0,5%C

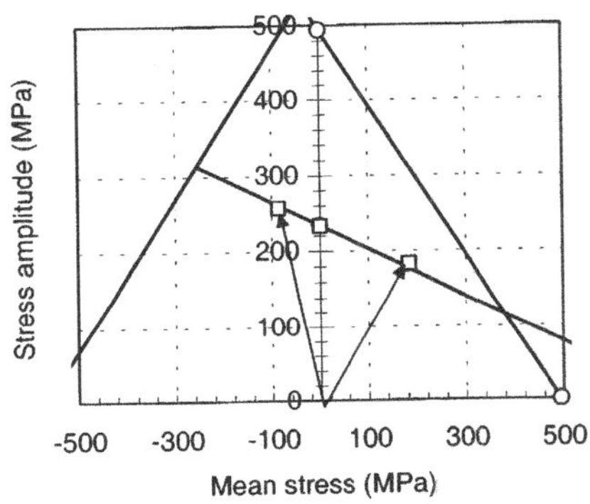

Figure 2. Haig diagram showing influence of load ratio on the allowable stress amplitude. Material: Distaloy AE + 0,5%C

The fatigue limit is strongly dependent on density, see Figure 3. Furthermore, the yield and ultimate tensile stresses have also a strong influence on the fatigue limit. Surface densification can increase the fatigue limit significantly, specially in the case of bending, see Figure 4.

THE FATIGUE PROCESS

The fatigue process can be divided into three phases; crack initiation, crack growth and final failure. The crack initiation occurs at some undefined time between the start of the test and the failure. If the load is below approximately 180 MPa, in Figure 1, the does not initiate or if it does it never grows to a length enough to cause failure. The S-N diagram can not give us information on the relation between the life spent without cracks and the duration of the crack propagation phase. However, it is well suited for the design of a component. For applications requiring many cycles to failure, the mean and scatter of the fatigue limit can be used to define a load that is safe with a chosen probability. If the load is varying to a significant degree a cumulative damage theory can be used in order to utilise some of the finite life regime too.

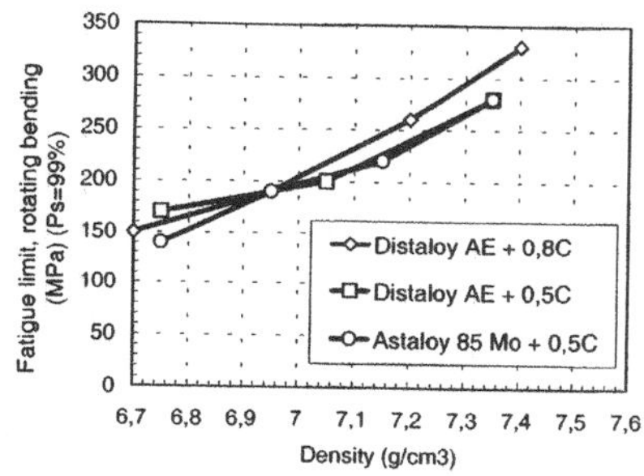

Figure 3. Fatigue limit - density diagram, showing difference between axial- and bend fatigue test.

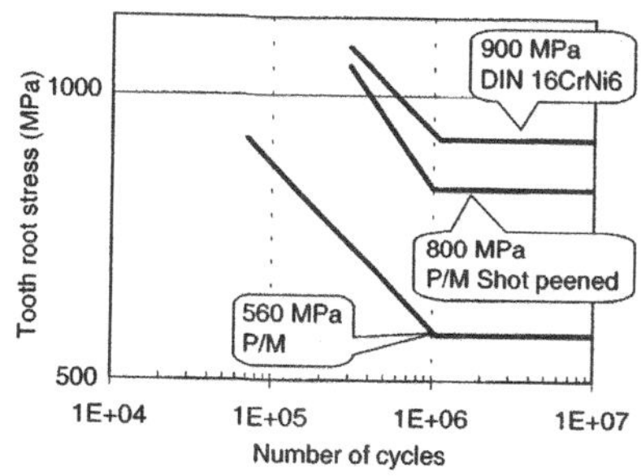

Figure 4. S–N diagram showing influence of surface densification on the fatigue limit. Material: Astaloy Mo + 0,2%C + case hardening

The crack initiation of high-density P/M steel (>7,3 g/cm^3) is controlled by defects in the microstructure. These defects can be pore clusters or inclusions, usually not individual pores since they are too small at this density. The population of large pores and the ability of the microstructure to resist initiation mainly govern the crack initiation in medium-density P/M steel (6,8 - 7,3 g/cm3). Pre-mixed and diffusion alloyed P/M steels are superior to pre-alloyed steels, simply because the sinter necks are stronger due to higher alloying content. Low-density P/M steels (<6,8 g/cm^3) generally exhibit early crack initiation.

Crack growth in high-density P/M steels is controlled by the microstructure. The bulk of the material must be strong enough to resist trans-granular fracture and the sinter necks must be strong and ductile enough to resist inter granular fracture. If both these conditions are present the axial fatigue limit can reach 250 MPa in the as-sintered state and 280 MPa in the tempered condition. Crack propagation is slower than for wrought steel near the threshold limit, then faster. The same situation is true for the medium density P/M steels, but at a lower level of course. Furthermore, the pores can cause crack deflection, thus delaying the fatigue life. For low-density P/M steels the crack propagation rate is governed by the strength of the sintering necks. Mixed and diffusion alloyed P/M steels are superior to pure iron or pre-alloyed P/M steel.

The resistance to fracture as expressed as fracture toughness or impact strength is low compared to wrought steel. The resistance to fracture in high- and medium-density P/M steels depends on the microstructure; hard structures give cleavage fracture and softer structure gives dimples. However, due to the constraint on deformation imposed by the pores, even the softer structures exhibit a low strain to fracture. In the low-density P/M steel all the deformation is concentrated to the sintering necks. This means that a small volume of material is involved in the fracture.

DESIGN CONSIDERATIONS

The first thing to consider is if the component is suitable for manufacturing by the P/M route at all. The specific requirements for the design of P/M components will not be taken up here. The requirements for P/M components are in general the same as for wrought steel, i.e. the surface finish should be high, stress concentrations should be minimised, etc. However, it should be pointed out that the surface roughness of a P/M component usually is quite good. Burr should be avoided or removed since (large) burr decreases the fatigue limit. In a component where notches can not be avoided, P/M steels are a competitive selection due to the low notch sensitivity. The ability of the process to achieve a minimum density should be given careful thought. Density variations are always present in a pressed body to some degree. If the geometry and the press tool are unfavourable, the variation can be as much as 0,5 g/cm^3 from one position to another. By changing the design, or more likely, selecting or designing an improved press tool a better density distribution can be achieved.

The information that is needed in order to be able to select material or optimise the component design is listed below:

Fatigue limit or endurance limit (50%)	The probability of survival (Ps) must be known, but not necessarily 50%.
Standard deviation of fatigue limit	If the fatigue limit is known at two Ps levels, the standard deviation can be calculated from these numbers.
Number of cycles for run outs e.g. 2 10^6, 5 10^6, 10^7	Certain applications require a very high number of load cycles, e.g. conrod. If the material does not have a marked yield point (true for most P/M materials) the fatigue limit determined at 10^8 may be lower than at 2 10^6.
Notch sensitivity	S-N diagram of a test with a notch factor in the vicinity of the current one
Type of testing: axial, bending, etc.	As close to the current load situation as possible.
Load ratio: R=-1, R=0,	As close to the current load situation as possible. Haigh diagram

PROTOTYPES

Prototypes can be manufactured from blanks. The fatigue properties of the prototypes are very near the properties of the pressed component. The deviations that occur depends on two things; the surface condition of the prototype and the density variations within the pressed component. Milling, turning, drilling are operations that can be performed with a wide span of cutting parameters. Wire erosion can be used, but the surface finish must correspond to the pressed component and shot peening should not be performed in order to remove the black scales from the erosion. In some cases it is possible to manufacture the prototypes in the green state (before sintering). Care must be taken not to cause cracks in the fragile green body during clamping and cutting. The density variation in the P/M blanks is usually very low compared to as-pressed components. This means that the minimum density achievable in the component must be considered.

MICROSTRUCTURE, PORES, ALLOYING SYSTEMS

The microstructure of P/M steels can be divided in two groups; the inhomogeneous materials manufactured by mixing iron and alloy powders or using diffusion alloyed powder and press and sinter. In these components the alloying elements are located between the iron particles. The other group of microstructures is the pre-alloyed powder. These microstructures are homogeneous in terms of alloying elements. Usually the sinter necks are stronger for the first type (high alloy content) and the grains (bulk) for the second. The P/M material most resistant to fatigue consist of pre-alloyed powder (e.g. 1,5 Mo) and Ni and Cu added as a pre-mix/diffusion alloy (e.g. Distaloy HP-1). Admixed nickel is the superior choice for increasing the fatigue limit. Warm compaction provides a beneficial pore structure resulting in improved fatigue properties.

SECONDARY OPERATIONS

Case hardening, plasma nitriding, carbo nitriding and other surface hardening methods are very beneficial for the fatigue properties. These methods not only increase the strength of the surface layer, they also induce compressive stresses in it. For bend loading this is the ideal situation and increases in fatigue limit of a factor of two can be achieved. Uniform heat-treatments such as; tempering and quench and temper, increases the yield stress and therefore increases the fatigue limit. Surface densification can be achieved by rolling, sizing or shot peening. The latter two methods are used without heat treatment, thus utilising the compressive stresses induced by the surface deformation, see Figure 4. Surface rolling is used for gears and they usually required a case hardening after the rolling, thus equilibrating the compressive stresses induced by the deformation.

S649/006/99

Using powder metallurgy (PM) for friction materials

J KETTELL
Wellman Friction Products, Sheffield, UK

1. INTRODUCTION

Material options, process methods, applications and installation requirements are detailed in this paper. Friction is basically a black art due to the infinite variety of parameters involved in establishing performance requirements for a particular installation.

Sintering is similar to baking a cake, by taking various powders, mixing and moulding into the required shape and then heating in an oven. The finished product is a structure formed by the various powders being fused together. Where grains of powder touch they fuse together in varying degrees, depending on the various temperature characteristics of the powders, thus producing a honey comb structure.

Process temperature is critical to ensure that the structure is correctly sintered. Too high a temperature will result in a structure fused into a solid mass and at the other end of the scale, too low a temperature will result in the powder falling apart when handled, machined or worked on.

2. POWDER FORMULA AND PROCESS ROUTES

Powder formulation will depend on the performance requirements for the application. The range is very wide, but will follow fairly similar patterns and this is where manufacturers generally have a list of standard materials available to suit specific applications. Development of new material formulations continues.

Typical ingredients are copper and tin with balancing additives such as ceramics, graphite and iron powders.

There are two main process routes used for sintering friction materials onto the steel plates. The basic process is the same for most sintering which involves powder mix, compacting, sintering and machining. There are variations depending on the process used where the compacting is done after sintering and machining is replaced by pressing grooves and pressing thickness dimensions to size.

Figure 1/1 shows the pre-compacting process route.

The powder mix is formed in a die into a "green" compact ring or pad shape to a specific density. These soft fragile rings are then applied to the steel core plate and the pre-assembly placed on a stack of other plates, placed under load in a "bell" furnace for the sintering process. Sintering temperatures for bronze materials are around 825 degrees centigrade. The sinter surfaces are then ground to size and any groove patterns machined into the surface.

Figure 1/2 shows the sprinkle route as used by Sundwiger Messingwerk.

The powder mix is in a hopper at the start of the "belt" furnace. The powder is "sprinkled" onto the steel core plate whilst on the moving belt, the plate then passing through the furnace. After sintering one side, the plate is returned to the beginning of the process to sinter a facing to the other side. In some applications, only one side has a sinter coating. This relatively thick layer of sintered coating now requires the structure reducing to the required size and density. This is carried out in a press and in addition to sizing the overall plate thickness, any groove patterns can also be pressed in, thereby eliminating any machining operations.

Where friction material is not required in specific areas of the steel plate, masks are used whilst the powder is applied to the steel plate and lifted off prior to passing through the belt furnace. Flatness is obtained in a separate heat flattening process.

The end result is the same, producing a powder structure which has a degree of porosity, frictional value and wear capability depending on application requirements.

Figures 2 & 3 show two micro structures of a typical sinter formulation. Figure 2 shows a higher density compared to figure 3 which is of a lower density.

3. FRICTION VALUES

Friction values may range anywhere between a low a 0.05 up to 0.50 mu depending on powder mix, dry or oil filled applications, density and performance requirements.

Friction performance will therefore depend on a variety of parameters relative to the application.

Plate diameter, rubbing speed, oil or dry conditions, loading, energy dissipation and temperature performance will effect actual friction values over the operating range. Static and Dynamic friction values also need to be considered when selecting a material for an application.

See figure 4 showing a typical friction curve.

4. DRY OR OIL OPERATION

Porosity/density of the sinter material can be effectively used as a bearing because of its capability in holding oil within its structure. It is this characteristic which provides the oil cooling potential in axles and transmissions.

The other feature of sinter bronze is its capability to work dry with good performance and durability potential.

Oil filled applications generally rely on the plates dipping in the oil and hope that the oil gets to where it will help with the friction level and to absorb the heat generated. The preferred method is to provide a pumped system where the oil is metered across the friction surfaces to provide adequate cooling and the best performance for operation and durability.

5. GROOVE PATTERNS

Sinter materials working dry do not necessarily require grooves in the face, but the introduction of a basic groove pattern can help to get rid of wear debris from between the friction surfaces.

Where the application is in an oil mist or is oil immersed, it is necessary to have grooves in the sinter face to break the oil film, otherwise engagement of the brake or clutch will be very slow or may not engage at all. Typically depending on oil type and application loading, the oil film has to be squeezed and wiped from between the friction faces and may be slow in break down and will act like a bearing until the surfaces actually contact.

Therefore oiled applications require some sort of groove pattern. Groove patterns and groove shape can be used to tune the friction performance. Depending on the application, the groove shape and pattern can used to meter oil across the faces and this can affect the friction value and performance of the application. The other problem of noise generated by the brake or clutch can be reduced with special attention to the groove patterns and also the oil type.

See figure 5 for typical groove patterns.

Groove shape also has an effect on friction performance relative to the wiping effect in clearing the oil from between the friction surfaces. The groove shape and pattern effects oil flow across the mating surfaces and therefore the working temperature and performance.

See figure 6 for typical groove shapes and the effect on wear for a given set of parameters.

6. OPERATIONAL DETAILS

Sintered frictional material for oil applications depends on the oil for its frictional characteristics. Oils within transmissions and braking systems can contain friction modifiers and these can drastically effect the friction value through the operating range.

The nature of sinter material is its porous capability within its structure which can be used for the benefit of a large range of applications. Sinter bronze friction material can be used in dry conditions, oil filled applications and also in exposed conditions where they are open to the elements. Organic materials will suffer problems of slurry development when exposed to damp or water conditions, resulting a drop in the friction value whereas sinter will disperse the water as if it was oil.

Medium to high carbon steel is used for the core plate to support the sinter. A key or flux is required to ensure a good bond between the steel and the sinter material. Without this, the sinter would just peel off the steel plate

Varying degrees of energy absorption will need to be considered for clutch and brake applications, the range of solutions is infinite when considering the selection of materials and defining the parameters. Attention must be made to the heat absorption capability of the mating surfaces which should be of carbon steel or similar properties with a minimum hardness of 220 Brinell. Sinter materials contain both abrasives and sold lubricants within the powder mix which may cause softer mating materials to wear excessively.

The basic thing to remember is that every application is different and the solutions will vary depending performance requirements.

7. TYPICAL APPLICATIONS

CLUTCHES & BRAKES	In single plate or multiplate format. Oiled or dry.
POWER SHIFT TRANSMISSIONS	Normally multiplate
TORQUE LIMITING CLUTCHES	
LIMITED SLIP DIFFERENTIAL PLATES	
AIRCRAFT BRAKES	
DISC BRAKE PADS	
SAFETY BRAKES	
BRAKE STRIPS AND PADS FOR LEISURE PARK RIDES	

FIGURE 1/1

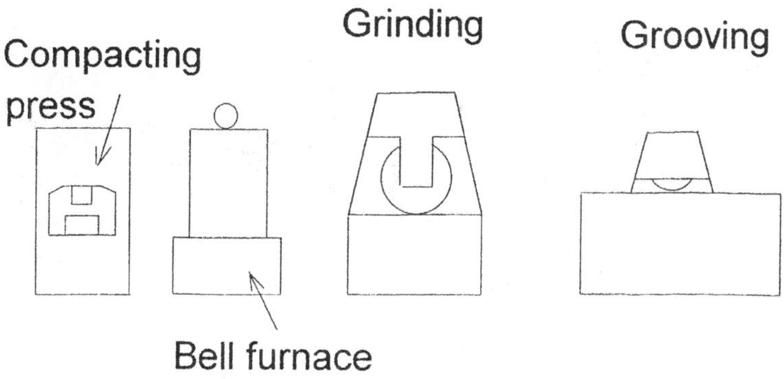

FIGURE 1/2

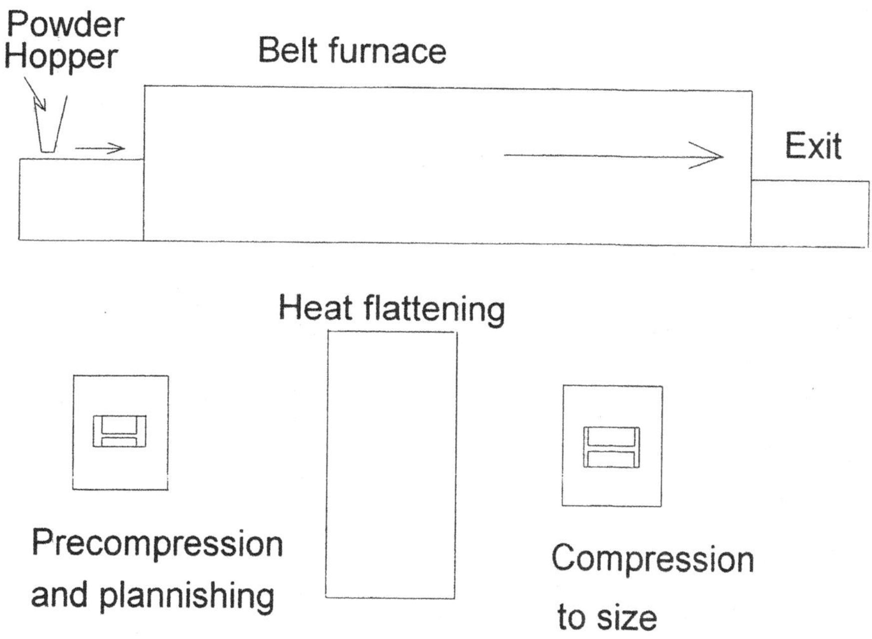

FIGURE 2

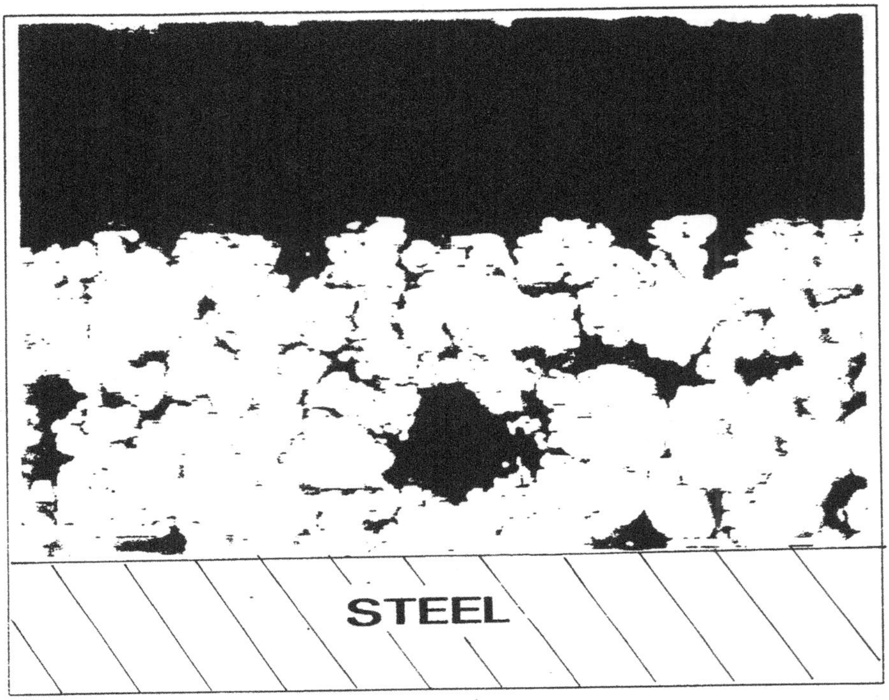

Vergrößerung ca. 700:1

Higher density - volume of pores approximately 25%

FIGURE 3

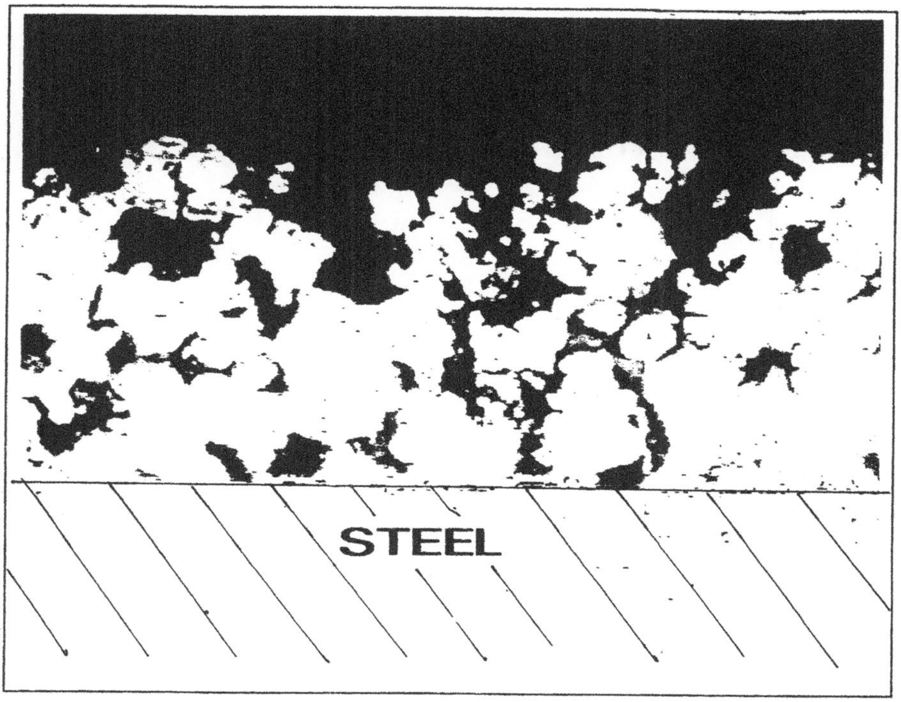

Vergrößerung ca. 700 : 1

Lower density - volume of pores approximately 40%

FIGURE 4

TYPICAL FRICTION ENGAGEMENT GRAPH

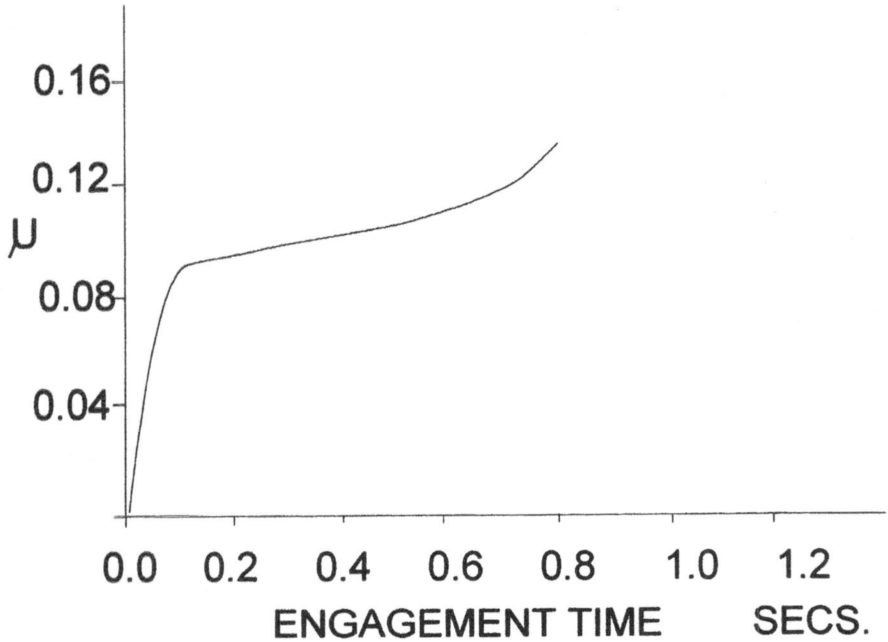

Oil application, 6 friction surfaces

The ratio of static to dynamic is 1.40:1

from first engagement to lock-up

TYPICAL GROOVE PATTERNS

FIGURE 5

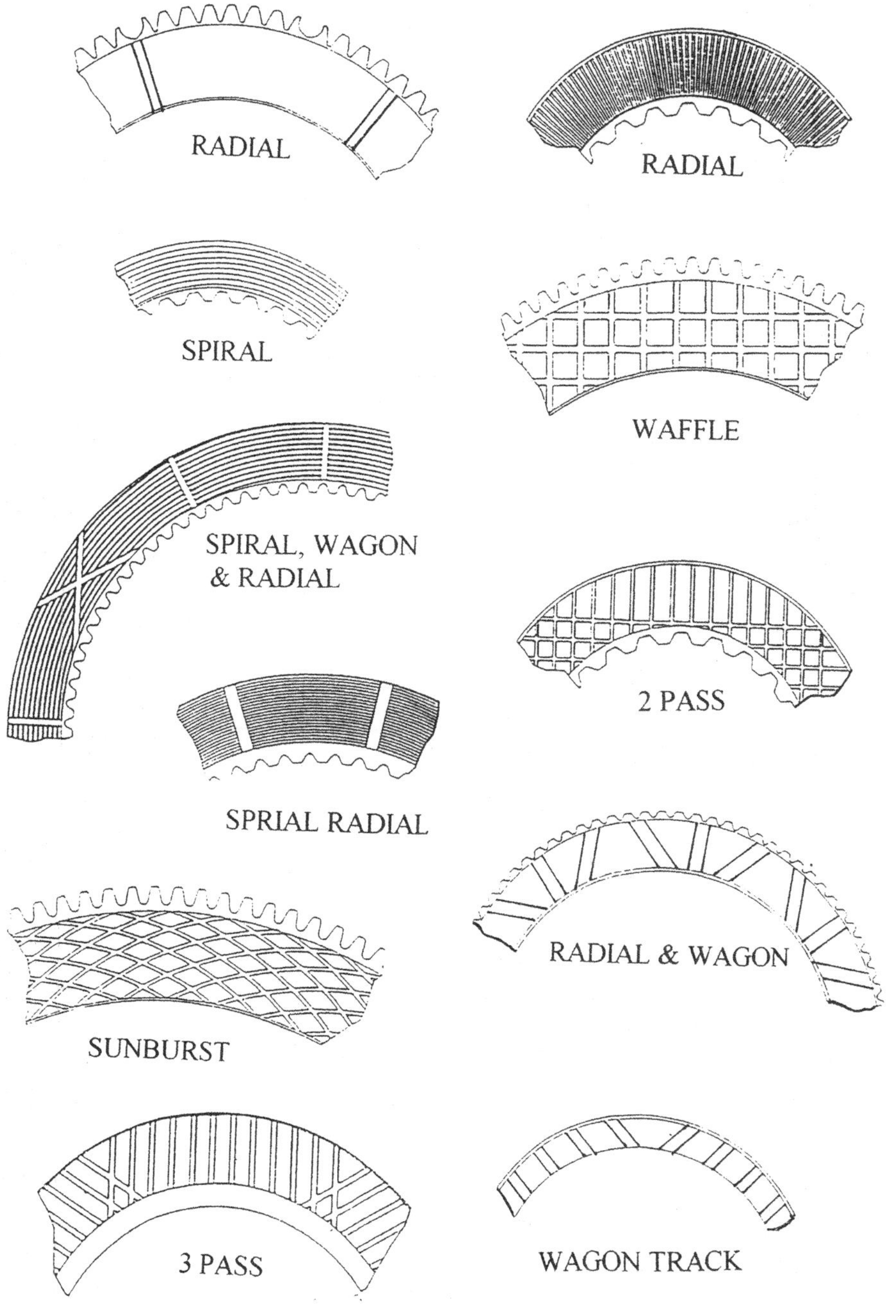

FIGURE 6

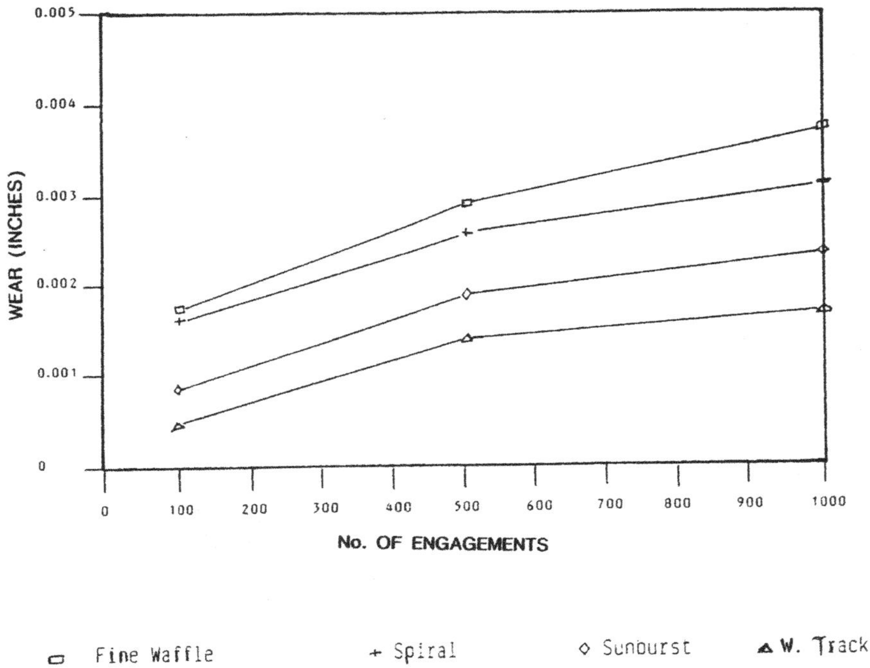

Figure III

DIFFERENT GROOVE SHAPES

Printed and bound by CPI Group (UK) Ltd, Croydon, CR0 4YY

13/09/2026

14961032-0002